MATHEMATICAL REPORTS

ARTHUR ZIFFER

AuthorHouse™ LLC
1663 Liberty Drive
Bloomington, IN 47403
www.authorhouse.com
Phone: 1-800-839-8640

Published by AuthorHouse 07/21/2014

ISBN: 978-1-4969-2843-6 (sc)
ISBN: 978-1-4969-2842-9 (e)

Library of Congress Control Number: 2014946630

Contents

The Discrete Kalman Filter

The Equations for the Discrete Kalman Filter

Ellen Hetland Fenwick
Arthur Jay Ziffer
February 25, 1988

Basic Concepts:

From the introduction to the IEEE collection of papers on the Kalman Filter the following is taken: "It is probably not an overstatement to assert that the Kalman Filter represents the most widely applied and demonstratably useful result to emerge from the state variable approach of 'modern control theory' ". The following talk will give an introduction to the five equations that can be considered to be the heart of the discrete Kalman Filter.

We start with the following equations:

$$\vec{X}_k = \Phi_{k-1}\vec{X}_{k-1} + \vec{W}_{k-1},$$

where $\vec{W}_k \epsilon N(0, Q_k)$; that is, $\vec{W}_k$ is a normally distributed random variable with mean, 0, and covariance, Q_k. Further

$$\vec{Z}_k = H_k\vec{X}_k + \vec{V}_k,$$

where $V_k \epsilon N(0, R_k)$; that is, $\vec{V}_k$ is a normally distributed random variable with mean, 0, and covariance, R_k.

These are preliminary to the five equations mentioned above. The first is called the system model; the second is called the measurement model. The $\vec{X}_k$ and $\vec{W}_k$ are $n \times 1$ vectors, and the $\vec{Z}_k$ and $\vec{V}_k$ are $m \times 1$ vectors. Φ and Q_k are $n \times n$ matrices, H_k is an $m \times n$ matrix and R_k is an $m \times m$ matrix. The subscript, k, refers to the discrete time, t_k. The following identifications are understood:

$$\begin{aligned}
X_k &= X(t_k),\\
W_k &= W(t_k),\\
Z_k &= Z(t_k),\\
V_k &= V(t_k),\\
\Phi_k &= \Phi(t_{k+1}, t_k),\\
Q_k &= Q(t_k),\\
H_k &= H(t_k);\\
R_k &= R(t_k).
\end{aligned}$$

Using the usual notation, $E(x)$, for the expectation of a random variable, x, the fact that $\vec{W}_k$ and $\vec{V}_k$ have mean zero and covariance matrices Q_k and R_k, respectively, are indicated by the following equations:

$$E[\vec{W}_k] = 0,$$
$$E[\vec{W}_k \vec{W}_k^T] = Q_k,$$
$$E[\vec{V}_k] = 0,$$
$$E[\vec{V}_k \vec{V}_k^T] = R_k.$$

Also, $\vec{W}_k$ and $\vec{V}_k$ are uncorrelated. That is,

$$E[\vec{W}_k \vec{V}_k^T] = 0,$$

for all k.

The Problem To Be Solved:

The problem that is to be solved through the use of the discrete Kalman Filter is to obtain estimates, $\hat{X}_k$, of $\vec{X}_k$ when we are given the measurement values, $\vec{Z}_k$. The following definitions are used for the estimated error, $\tilde{X}_k$, and the estimated covariance, P_k:

$$\tilde{X}_k = \hat{X}_k - \vec{X}_k,$$
$$P_k = E[\tilde{X}_k \tilde{X}_k^T].$$

Some Additional Assumptions:

The following will be needed in what follows:

$$E[\tilde{X}_k \vec{W}_k^T] = 0,$$
$$E[\vec{W}_k \tilde{X}_k^T] = 0,$$
$$E[\tilde{X}_k \vec{V}_k^T] = 0,$$
$$E[\vec{V}_k \tilde{X}_k^T] = 0.$$

The first set of relations results from $\vec{W}_k$ being white noise and the fact that system process noise and the estimation error at t_k are uncorrelated. The second set of relations results from $\vec{V}_k$ also being white noise and that measurement errors and estimation errors are uncorrelated. As initial conditions we have an initial estimate, $\hat{X}(0)$, of the state vector and the initial estimate of the covariance matrix, P_0, where

$$P_0 = E[\tilde{X}(0)\tilde{X}^T(0)].$$

The Five Kalman Filter Equations:

There are five equations considered to be the heart of the discrete Kalman Filter. Equations (1) and (2) are the time update (extrapolation) equations. Equations (3) and (5) are the measurement update equations. (4) is the equation for the Kalman gain matrix. We list the equations below:

$$\hat{X}_k(-) = \Phi_{k-1}\hat{X}_{k-1}(+), \tag{1}$$

$$P_k(-) = \Phi_{k-1}P_{k-1}(+)\Phi_{k-1}^T + Q_{k-1}, \tag{2}$$

$$\hat{X}_k(+) = \hat{X}_k(-) + K_k\left(\vec{Z}_k - H_k\hat{X}_k(-)\right), \tag{3}$$

$$K_k = P_k(-)H_k^T\left(H_kP_k(-)H_k^T + R_k\right)^{-1}, \tag{4}$$

$$P_k(+) = (I - K_kH_k)\,P_k(-). \tag{5.}$$

Diagram I is the "Timing Diagram" of the various quantities involved above.

Equations (1) and (2) describe the behavior of the state estimate, $\vec{X}_k$, and error covariance, P_k, between measurements. Equations (3) and (5) with the use of (4) describe the state estimate and error covariance behavior across a measurement.

The Derivation of Equation (1):

To derive equation (1) we start with the system model equation

$$\vec{X}_k = \Phi_{k-1}\vec{X}_{k-1} + \vec{W}_{k-1}.$$

Taking expectations, we get

$$\begin{aligned}\hat{X}_k &= E[\vec{X}_k],\\ &= E[\Phi_{k-1}\vec{X}_{k-1} + \vec{W}_{k-1}],\\ &= \Phi_{k-1}E[\vec{X}_{k-1}] + E[\vec{W}_{k-1}],\\ &= \Phi_{k-1}\hat{X}_{k-1},\end{aligned}$$

since $\vec{W}_k$ has zero mean (expectation). From the relation,

$$\hat{X}_k = \Phi_{k-1}\hat{X}_{k-1},$$

we can write

$$\hat{X}_k(-) = \Phi_{k-1}\hat{X}_{k-1}(+),$$

where $\hat{X}_k(-)$ is the a priori estimate immediately before the discrete measurement, $\vec{Z}_k$ is used and $\hat{X}_{k-1}(+)$ is the a posteriori update after the measurement, $\vec{Z}_{k-1}$, is used.

The derivation of Equation (2):

To derive (2), we proceed as follows:

$$\begin{aligned}
\tilde{X}_{k+1} &= \hat{X}_{k+1} - \vec{X}_{k+1}, \\
&= \Phi_k\hat{X}_k - \left(\Phi_k\vec{X}_k - \vec{W}_k\right), \\
&= \Phi_k\left(\hat{X}_k - \vec{X}_k\right) - \vec{W}_k, \\
&= \Phi_k\tilde{X}_k - \vec{W}_k.
\end{aligned}$$

Since

$$\begin{aligned}
P_{k+1} &= E[\tilde{X}_{k+1}\tilde{X}_{k+1}^T], \\
&= E[\left(\Phi_k\tilde{X}_k - \vec{W}_k\right)\left(\Phi_k\tilde{X}_k - \vec{W}_k\right)^T], \\
&= E[\left(\Phi_k\tilde{X}_k - \vec{W}_k\right)\left(\tilde{X}_k^T\Phi_k^T - \vec{W}_k^T\right)], \\
&= E[\Phi_k\tilde{X}_k\tilde{X}_k^T\Phi_k^T - \Phi_k\tilde{X}_k\vec{W}_k^T - \vec{W}_k\tilde{X}_k^T\Phi_k^T + \vec{W}_k\vec{W}_k^T], \\
&= \Phi_k E[\tilde{X}_k\tilde{X}_k^T]\Phi_k^T - \Phi_k E[\tilde{X}_k\vec{W}_k^T] - E[\vec{W}_k\tilde{X}_k^T]\Phi_k^T + E[\vec{W}_k\vec{W}_k^T], \\
&= \Phi_k P_k\Phi_k^T + Q_k.
\end{aligned}$$

Now we invoke (+) and (-) considerations as we did to derive equation (1) and we get

$$P_k(-) = \Phi_{k-1}P_{k-1}(+)\Phi_{k-1}^T + Q_{k-1}.$$

Some Interim Remarks:

Thus we have derived the time propagation equations. We shall now derive the measurement update equations and the formula for the Kalman gain matrix. That is, given an a priori estimate of the system state, $\hat{X}_k(-)$, at time t_k, we seek an updated estimate, $\hat{X}_k(+)$, which uses the measurement $\vec{Z}_k$. In order to avoid an evergrowing memory filter, this updated estimate is sought by means of the linear recursive form,

$$\hat{X}_k(+) = L_k\hat{X}_k(-) + K_kZ_k,$$

where L_k and K_k are $n \times n$ and $n \times m$ matrices, respectively.

The Derivation of Equation (3):

From the relations

$$\hat{X}_k(+) = \vec{X}_k + \tilde{X}_k(+)$$

and

$$\hat{X}_k(-) = \vec{X}_k + \tilde{X}_k(-)$$

and the proposed recursive form above we have the following development:

$$\begin{aligned}
\tilde{X}_k(+) &= \hat{X}_k(+) + X_k, \\
&= \left(L_k \hat{X}_k(-) + K_k Z_k\right) - \vec{X}_k, \\
&= L_k \left(\vec{X}_k + \tilde{X}_k(-)\right) + K_k \left(H_k \vec{X}_k + \vec{V}_k\right) - \vec{X}_k, \\
&= (L_k + K_k H_k - I))\, \vec{X}_k + L_k \tilde{X}_k(-) + K_k \vec{V}_k.
\end{aligned}$$

Taking the expected value of both sides of

$$\tilde{X}_k(+) = (L_k + K_k H_k - I)\, \vec{X}_k + L_k \tilde{X}_k(-) + K_k \vec{V}_k$$

yields

$$E[\tilde{X}_k(+) = (L_k + K_k H_k - I)\, E[\vec{X}_k] + L_k E[\tilde{X}_k(-)] + K_k E[\vec{V}_k].$$

Since the mean of $\vec{V}_k$ is zero, $E[\vec{V}_k] = 0$. Furthermore, if $\tilde{X}_k(-)$ is an unbiased estimate (that is, $E[\tilde{X}_k(-) = 0$) then the estimator $\tilde{X}_k(+)$ will be unbiased (that is, $E[\tilde{X}_k(+) = 0$) for any given state vector, $\vec{X}_k$, only if

$$L_k + K_k H_k - I = 0,$$

or

$$L_k = I - K_k H_k.$$

Hence the expression,

$$\hat{X}_k(+) = L_k\hat{X}_k(-) + K_k\vec{Z}_k$$

can be written

$$\hat{X}_k(+) = (I - K_kH_k)\,\hat{X}_k(-) + K_k\vec{Z}_k.$$

By rearranging, we get

$$\hat{X}_k(+) = \hat{X}_k(-) + K_k[\vec{Z}_k - H_k\hat{X}_k(-)].$$

The Derivation of Equation (4):

The corresponding estimation error is derived as follows:

$$\begin{aligned}
\tilde{X}_k(+) &= (L_k + K_kH_k - I)\,\vec{X}_k + L_k\tilde{X}_k(-) + K_k\vec{V}_k,\\
&= (L_k + K_kH_k - I)\left(\hat{X}_k(-) - \tilde{X}_k(-)\right) + L_k\tilde{X}_k(-) + K_k\vec{v}_k,\\
&= (l_k + K_kH_k - I)\,\hat{X}_k(-) + (I - K_kH_k - L_k)\,\tilde{X}_k(-) + L_k\tilde{X}_k(-) + K_k\vec{V}_k,\\
&= (0)\hat{X}_k(-) + (I - K_kH_k)\,\tilde{X}_k(-) - L_k\tilde{X}_k(-) + L_k\tilde{X}_k(-) + K_k\vec{V}_k,\\
&= (I - K_kH_k)\,\tilde{X}_k(-) + K_k\vec{V}_k.
\end{aligned}$$

Hence

$$\begin{aligned}
P_k(+) &= E[\tilde{X}_k(+)\tilde{X}_k^T(+)],\\
&= E[(I - K_kH_k)\,\tilde{X}_k(-)\left(\tilde{X}_k^T(-)\,(I - K_kH_k)^T + \vec{V}_k^T K_k^T\right)\\
&\quad + K_k\vec{V}_k\left(\tilde{X}_k^T(-)\,(I - K_kH_k)^T + \vec{V}_k^T K_k^T\right)],\\
&= (I - K_kH_k)\left(E[\tilde{X}_k(-)\tilde{X}_k^T(-)]\,(I - K_kH_k)^T + E[\tilde{X}_k(-)\vec{V}_k^T]K_k^T\right)\\
&\quad + K_k\left(E[\vec{V}_k\tilde{X}_k^T(-)]\,(I - K_kH_k)^T + E[\vec{V}_k\vec{V}_k^T]\right)K_k^T.
\end{aligned}$$

By the use of

$$E[\tilde{X}_k(-)\vec{V}_k^T] = E[\vec{V}_k\tilde{X}_k^T(-)] = 0,$$

we arrive at

$$P_k = (I - K_kH_k)\,P_k(-)\,(I - K_kH_k)^T + K_kR_kK_k^T.$$

The criterion for choosing K_k is to minimize a weighted scalar sum of the diagonal elements of the error covariance matrix, $P_k(+)$:

$$J_k = E[\tilde{X}_k^T(-) S \tilde{X}_k(+)],$$

where S is any positive semidefinite matrix. The optimal estimate can be shown can be shown to be independent of S, hence we let $S = I$, where I is the identity matrix; thus J_k can be written as follows:

$$\begin{aligned}
J_k &= E[\tilde{X}_k^T(+) I \tilde{X}_k(+)], \\
&= E[\sum_{i=1}^{i=n} \tilde{X}_{k_i}^2(+)], \\
&= \sum_{i=1}^{i=n} E[X_{k_i}^2(+)], \\
&= trace[P_k(+)], \\
&= tr[P_k(+)].
\end{aligned}$$

This is equivalent to minimizing the length of the estimation error vector. To find the value of K_k which provides a minimum, it is necessary to take the partial derivatives of J_k with respect to K_k and equate them to zero. We recall, first, some theorems from matrix differential calculus for matrices A, B, C and D and a symmetric matrix, B:

$$\frac{\partial}{\partial A}\left(tr\left(ABA^T\right)\right) = 2AB,$$

$$\frac{\partial}{\partial A}\left(tr\left(BAC\right)\right) = B_T C_T,$$

where the definition for the derivative of a scalar, z with respect to a matrix, $A = (a_{ij})$ is given by

$$\frac{\partial z}{\partial A} = \left(\frac{\partial z}{\partial a_{ij}}\right).$$

From the relation

$$P_k(+) = (I - K_k H_k) P_k(-) (I - K_k H_k)^T + K_k R_k K_k^T,$$

we arrive at

$$P_k = P_k(-) - K_k H_k P_k(-) - P_k H_k^T K_k^T + K_k H_k P_k(-) H_k^T K_k^T + K_k R K_k^T.$$

Using the relations

$$tr(A+B) = tr(A) + tr(B),$$
$$tr\left(A^T\right) = tr(A),$$

and the fact that estimated covariances matrices are symmetric we can write

$$tr(P_k(+)) = tr(P_k(+)) - 2tr(K_k H_k P_k(+)) + tr\left(K_k H_k P_k H_k^T(+) K_k\right) + tr\left(K_k R_k K_k^T\right).$$

Taking derivatives with respect to K of $tr(P_k(+))$, we get

$$\frac{\partial(tr(P_k(+)))}{\partial K_k} = 0 - 2P_k(+)H_k^T + 2K_k H_k P_k(+)H_k^T + 2K_k R_k.$$

Setting the partial derivative equal to zero and solving for K yields

$$K_k = P_k(-)H_k^T\left(H_k P_k(-)H_k^T + R_k\right)^{-1},$$

which is equation (4).

The Derivation of Equation (5):

To obtain equation (5) we substitute the equation

$$K_k = P_k(-)H_k^T\left(H_k P_k(-)H_k^T + R_k\right)^{-1}$$

into

$$P_k(+) = P_k(+) - K_k H_k P_k(+) - P_k(+)H_k^T K_k^T + K_k H_k P_k(+)H_k^T K_k^T + K_k R_k K_k^T.$$

For ease of reading, let us make the substitution

$$\alpha = \left(H_k P_k(-)H_k^T + R_k\right)^{-1}.$$

Then we have

$$\begin{aligned}
P_k(+) &= P, \\
&= P - PH^T\alpha HP - PH^T\alpha^T HP + PH^T\alpha HPH^T\alpha^T HP + PH^T\alpha R\alpha^T HP, \\
&= P - PH^T\alpha HP - PH^T\alpha^T HP + PH^T\alpha\alpha^{-1}\alpha^T HP, \\
&= P - PH^T\alpha HP - PH^T\alpha^T HP + PH^T\alpha^{-1}\alpha^T HP, \\
&= P - PH^T\alpha HP - PH^T\alpha^T HP + PH^T\alpha\alpha^{-1}\alpha^T HP, \\
&= P - PH^T\alpha HP - PH^T\alpha^T HP + PH^T\alpha^T HP, \\
&= P - PH^T\alpha HP, \\
&= P - KHP, \\
&= (I - KH)P, \\
&= (I - K_k H_k)P_k(-),
\end{aligned}$$

where we have dropped subscripts and the symbol, (–), in the body of this last calculation in order to make the calculation more clear. This is equation (5).

Summary of Equations
in Report:
The Discrete Kalman Filter
including Diagram I

SINGER

0.0

SUMMARY OF DISCRETE KALMAN FILTER EQUATIONS

System Model	$\underline{x}_k = \Phi_{k-1}\underline{x}_{k-1} + \underline{w}_{k-1}, \quad \underline{w}_k \sim N(\underline{0}, Q_k)$
Measurement Model	$\underline{z}_k = H_k\underline{x}_k + \underline{v}_k, \quad \underline{v}_k \sim N(\underline{0}, R_k)$
Initial Conditions	$E[\underline{x}(0)] = \hat{\underline{x}}_0, \; E[(\underline{x}(0) - \hat{\underline{x}}_0)(\underline{x}(0) - \hat{\underline{x}}_0)^T] = P_0$
Other Assumptions	$E[\underline{w}_k\underline{v}_j^T] = 0$ for all j, k
State Estimate Extrapolation	$\hat{\underline{x}}_k(-) = \Phi_{k-1}\hat{\underline{x}}_{k-1}(+)$
Error Covariance Extrapolation	$P_k(-) = \Phi_{k-1}P_{k-1}(+)\Phi_{k-1}^T + Q_{k-1}$
State Estimate Update	$\hat{\underline{x}}_k(+) = \hat{\underline{x}}_k(-) + K_k[\underline{z}_k - H_k\hat{\underline{x}}_k(-)]$
Error Covariance Update	$P_k(+) = [I - K_kH_k]P_k(-)$
Kalman Gain Matrix	$K_k = P_k(-)H_k^T[H_kP_k(-)H_k^T + R_k]^{-1}$

THE SINGER COMPANY • ELECTRONIC SYSTEMS DIVISION

SINGER

JTIDS

$$x_k = \phi_{k-1} \, x_{k-1} + w_{k-1} \qquad w_k \sim N(0, q_k)$$

$$z_k = H_k \, x_k + v_k \qquad v_k \sim N(0, R_k)$$

$$x_k = x(t_k)$$

$$z_k = z(t_k)$$

$$\vdots$$

$$\phi_k = \phi(t_{k+1}, t_k)$$

THE SINGER COMPANY • ELECTRONIC SYSTEMS DIVISION

SINGER

DIAGRAM I

JTIDS

0.5

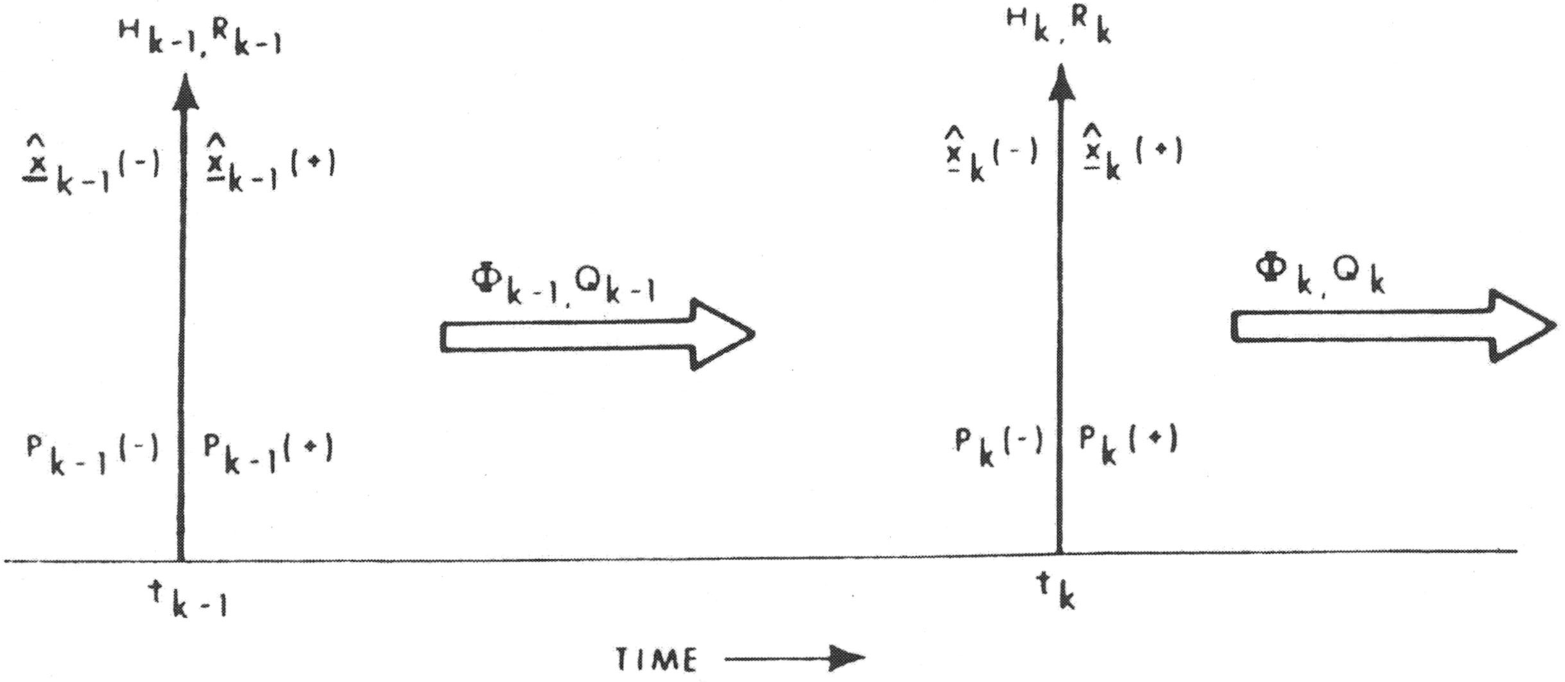

THE SINGER COMPANY • ELECTRONIC SYSTEMS DIVISION

$$E[w_k] = 0$$

$$E[w_k w_k^T] = Q_k$$

$$E[v_k] = 0$$

$$E[v_k v_k^T] = R_k$$

$$E[w_k v_j^T] = 0 \quad \text{for all } k \text{ and } j$$

SINGER

JTIDS

3

$$\tilde{x}_k \triangleq \hat{x}_k - x_k$$

$$P_k \triangleq E\left[\tilde{x}_k \tilde{x}_k^T\right]$$

$$E\left[\tilde{x}_k w_k^T\right] = E\left[w_k \tilde{x}_k^T\right] = 0$$

$$E\left[\tilde{x}_k v_k^T\right] = E\left[v_k \tilde{x}_k^T\right] = 0$$

INITIAL CONDITIONS: $\hat{x}(0)$, P_0

THE SINGER COMPANY • ELECTRONIC SYSTEMS DIVISION

JTIDS

4

$$x_k = \phi_{k-1} x_{k-1} + w_{k-1}$$

$$\hat{x}_k = E[x_k] = E[\phi_{k-1} x_{k-1} + w_{k-1}]$$

$$= \phi_{k-1} E[x_{k-1}] + E[w_{k-1}]$$

$$= \phi_{k-1} \hat{x}_{k-1}$$

(1) $$\hat{x}_k^{(-)} = \phi_{k-1} \hat{x}_{k-1}^{(+)}$$

SINGER

JTIDS

5

$$\tilde{x}_{k+1} \triangleq \hat{x}_{k+1} - x_{k+1}$$

$$= \phi_k \hat{x}_k - (\phi_k x_k + w_k)$$

$$= \phi_k (\hat{x}_k - x_k) - w_k$$

$$= \phi_k \tilde{x}_k - w_k$$

SINGER

JTIDS

6

$$P_{k+1} \triangleq E\left[\tilde{x}_{k+1}\tilde{x}_{k+1}^T\right]$$

$$= E\left[(\phi_k \tilde{x}_k - \omega_k)(\phi_k \tilde{x}_k - \omega_k)^T\right]$$

$$= E\left[(\phi_k \tilde{x}_k - \omega_k)(\tilde{x}_k^T \phi_k^T - \omega_k^T)\right]$$

$$= E\left[\phi \tilde{x}\tilde{x}^T\phi^T - \phi\tilde{x}\omega^T - \omega\tilde{x}^T\phi^T + \omega\omega^T\right]$$

$$= \phi E\left[\tilde{x}\tilde{x}^T\right]\phi^T - \phi E\left[\tilde{x}\omega^T\right] - E\left[\omega\tilde{x}^T\right]\phi^T + E\left[\omega\omega^T\right]$$

$$P_{k+1} = \phi_k P_k \phi_k^T + Q_k$$

(2) $$P_k(-) = \phi_{k-1} P_{k-1}^{(+)} \phi_{k-1}^T + Q_{k-1}$$

THE SINGER COMPANY • ELECTRONIC SYSTEMS DIVISION

$$\hat{x}_k(+) = L_k \hat{x}_k(-) + K_k z_k$$

$$\tilde{x}_k(+) = \hat{x}_k(+) - x_k$$

$$= (L_k \hat{x}_k(-) + K_k z_k) - x_k$$

$$= L_k (x_k + \tilde{x}_k(-)) + K_k (H_k x_k + v_k) - x_k$$

$$= (L_k + K_k H_k - I) x_k + L_k \tilde{x}_k(-) + K_k v_k$$

$$E[\tilde{x}_k(+)] = (L_k + K_k H_k - I) E[x_k] + L_k E[\tilde{x}_k(-)] + K_k E[v_k]$$

$$L_k = I - K_k H_k$$

(3) $$\hat{x}_k(+) = \hat{x}_k(-) + K_k [z_k - H_k \hat{x}_k(-)]$$

$$\tilde{x}_k(+) = (I - K_k H_k)\tilde{x}_k(-) + K_k v_k$$

$$P_k(+) = E\left[\tilde{x}_k(+)\,\tilde{x}_k(+)^T\right]$$

$$P_k(+) = E\left\{(I-KH)\tilde{x}\left[\tilde{x}^T(I-KH)^T + v^T K^T\right] + Kv\left[\tilde{x}^T(I-KH)^T + v^T K^T\right]\right\}$$

$$= (I-KH)\left\{E\left[\tilde{x}\tilde{x}^T\right](I-KH)^T + E\left[\tilde{x}v^T\right]K^T\right\}$$

$$+ K\left\{E\left[v\tilde{x}^T\right](I-KH)^T + E\left[vv^T\right]\right\}K^T$$

$$P_k(+) = (I - K_k H_k)P_k(-)(I - K_k H_k)^T + K_k R_k K_k^T$$

SINGER

8.5

Find K_k to minimize trace $[P_k(+)]$

$$P_k(+) = P - KHP - PH^TK^T + KHPH^TK^T + KRK^T$$

$$tr[P_k(+)] = tr[P] - 2\,tr[KHP] + tr[KHPH^TK^T] + tr[KRK^T]$$

$$\frac{\partial z}{\partial A} = \left(\frac{\partial z}{\partial a_{ij}}\right)$$

$$\frac{\partial\{tr[ABA^T]\}}{\partial A} = A(B+B^T), \quad \frac{\partial\{tr[BAC]\}}{\partial A} = B^TC^T$$

$$0 = \frac{\partial\{tr[P_k(+)]\}}{\partial K} = 0 - 2PH^T + 2KHPH^T + 2KR$$

$$K = PH^T[HPH^T + R]^{-1}$$

(4) $$K_k = P_k(-)H_k^T[H_kP_k(-)H_k^T + R_k]^{-1}$$

THE SINGER COMPANY • ELECTRONIC SYSTEMS DIVISION

SINGER

9

$$p = (1-kh)p'(1-kh) + krk$$

$$= (1-kh)^2 p' + k^2 r$$

$$= p' - 2p'hk + (p'h^2 + r)k^2$$

$$\frac{\partial p}{\partial k} = 0 - 2p'h + 2(p'h^2 + r)k$$

$$\frac{\partial p}{\partial k} = 0 \implies k = \frac{p'h}{h^2 p' + r} = \frac{p'h}{hp'h + r}$$

(4) $$K_k = P_k(-) H_k^T \left[H_k P_k(-) H_k^T + R_k \right]^{-1}$$

SINGER

JTIDS

10

$$P_k(+) = P - KHP - PH^TK^T + KHPH^TK^T + KRK^T$$

$$K = PH^T[\;\;]^{-1} \qquad P = P^T$$

$$P_k(+) = P - PH^T[\;\;]^{-1}HP - PH^T\{[\;\;]^{-1}\}^T HP$$

$$+ PH^T[\;\;]^{-1}HPH^T\{[\;\;]^{-1}\}^T HP$$

$$+ PH^T[\;\;]^{-1}R\{[\;\;]^{-1}\}^T HP$$

$$= P - PH^T[\;\;]^{-1}HP - PH^T\{[\;\;]^{-1}\}^T HP$$

$$+ PH^T[\;\;]^{-1}[\;\;]\{[\;\;]^{-1}\}^T HP$$

$$= P - PH^T[\;\;]^{-1}HP = P - KHP$$

(5) $$P_k(+) = P_k(-) - K_k H_k P_k(-)$$

Color Reports

Some Methods of Two-Constant Theory
Used in Determination of Colorant Mixtures

by

Ronald Jones
Ya Qi Li and
Arthur Ziffer

July 7, 1995

Introduction

For a mixture of n colorants, the absorption coefficient, K, and scattering coefficient, S, can be written as

$$K = \sum_{i=1}^{n} c_i k_i$$
$$S = \sum_{i=1}^{n} c_i s_i$$

where k_i and s_i are the absorption and scattering coefficients of colorant i, respectively, and c_i is the concentration of the i-th colorant. Hence

$$\frac{K}{S} = \frac{\sum_{i=1}^{n} c_i k_i}{\sum_{i=1}^{n} c_i s_i}$$

This is the expression for K/S of the colorant mixture in the two-constant theory (i.e. the k_i 's and the s_i 's appear separately). In the single-constant theory, the expression for K/S is

$$\frac{K}{S} = \sum_{i=1}^{n} c_i \left(\frac{k}{s}\right)_i$$

(i.e. the k_i 's and the s_i 's only appear as quotients $(k/s)_i$, that is they are combined and appear only as a single constant).

The purpose of calculating K and S is to determine the "color" of the mixture for a particular set of concentrations $(c_1, c_2, c_3, \ldots, c_n)$ of the n colorants in our proposed mixture. Once we have the values of K and S, we can calculate the reflectance of the mixture, R, from the Kubelka-Munk (KM) equation which in its general form gives R as a function of K, S, x, and R_g , where x is the film thickness of the mixture and R_g is the reflectance of the substrate; namely,

$$R = \frac{1 - R_g\,(a - b\,\text{coth}(bSx))}{a - R_g + b\,\text{coth}(bSx)}$$

$$a = 1 + K/S$$

$$b = (a^2 - 1)^{1/2}$$

$$(\text{thus } a^2 - b^2 = 1)$$

and coth is the hyperbolic cotangent. This, of course, is done for a discrete set of wavelengths from 400 nm to 700 nm (e.g. 400, 410, 420,...,700). The result of this is to obtain the reflectance curve of the mixture or rather the "color" of the mixture. If this is not a good "match" to some color sample we are trying to duplicate, then we change the c_i 's and redo the calculations until we are satisfied with the match. The more commonly used form of the KM equation is where the film thickness is considered infinite (x large). In this case, since R_g becomes 0 and coth(bSx) approaches 1 (the latter shown in the appendix), we have

$$R_\infty = \frac{1}{a+b} = \frac{a-b}{a^2 - b^2} = a - b = 1 + K/S - [(K/S)^2 + 2(K/S)]^{1/2}$$

Solving this equation for K/S gives the more common form of the KM equation

$$K/S = \frac{(1 - R_\infty)^2}{2\,R_\infty}$$

For later use we note that the general form of the KM equation can be solved for bcoth(bSx) and for Sx yielding

$$b\,\text{coth}(bSx) = \frac{1 + R\,R_g - a(R + R_g)}{R - R_g}$$

and

$$Sx = \frac{1}{b}\text{coth}^{-1}\{\frac{1 + R\,R_g - a\,(R + R_g)}{b(R - R_g)}\}$$

The above two equations are not usable when $R = R_g$ (no color).

Method I

Now we are ready to describe three methods for finding the k_i and s_i values for a particular colorant in a mixture (at a specified wavelength). The first method is from direct measurement of a contrast ratio. Consider a contrast ratio chart (that is half light (white) and half dark (black)) and draw down a film of the colorant under consideration over half the chart so that half the light and half the dark portions are covered by the colorant. If we assume that we have incomplete hide then we end up with a chart that has four different sections with reflectances R_{OL}, R_L , R_{OD} and R_D which are, respectively, reflectance over light, reflectance of light, reflectance over dark and reflectance of dark. From the form of the KM equation solved for $b\coth(bSx)$ above, we notice that all the measured reflectances are on the R.H.S. Hence if we take this equation and substitute $R = R_{OL}$ and $R_g = R_L$, we get

$$b\coth(bSx) = \frac{1 + R_{OL}\,R_L - a(R_{OL} + R_L)}{R_{OL} - R_L}$$

Then if we take the equation again, but this time substitute $R = R_{OD}$ and $R_g = R_D$, we get

$$b\coth(bSx) = \frac{1 + R_{OD}\,R_D - a(R_{OD} + R_D)}{R_{OD} - R_D}$$

Since the L.H.S of the above two equations are the same we can equate the R.H.S of both and get

$$\frac{1 + R_{OL}\,R_L - a(R_{OL} + R_L)}{R_{OL} - R_L} = \frac{1 + R_{OD}\,R_D - a(R_{OD} + R_D)}{R_{OD} - R_D}$$

Solving the above equation for a and recalling that $a = 1 + K/S$ we get the following for K/S

$$K/S = \frac{(1 + R_{OD}R_D)(R_L - R_{OL}) + (1 + R_{OL}R_L)(R_{OD} - R_D)}{2\,(R_{OD}R_L - R_{OL}R_D)} - 1$$

We note that in the opaque case $R_{OD} = R_{OL} = R_\infty$ and the R_D and R_L will cancel out of the above equation and give the expected form

$$K/S = \frac{(1 - R_\infty)^2}{2\,R_\infty}$$

An easy way to see this to let $R_{OD} = R_{OL} = R_\infty$ and let R_L and R_D take on the values 1 and 0, respectively, which are not inappropriate since we can identify light with white and dark with black. In either case we have K/S. From the form of the KM equation that gives an expression for Sx (previously given), we can find S if we know the value of x (film thickness). Once we have K/S and S we have K and we are finished (i.e. we have found k_i (= K above) and s_i (= S above) for this colorant).

Method II

The second method uses a mixture of a colorant and white. We use the standard term masstone for the colorant by itself. We have the expression

$$(K/S)_{mix} = \frac{K_{mix}}{S_{mix}} = \frac{c\,K + c_w\,K_w}{c\,S + c_w\,S_w}$$

where c, K, S are, respectively, the concentration, absorption coefficient and scattering coefficient of the colorant and c_w, K_w, S_w are analogously the concentration, absorption coefficient and scattering coefficient of the white being used. Replacing

$$K \text{ by } S(K/S)_{mass}$$

in the above equation gives

$$(K/S)_{mix} = \frac{c\,S\,(K/S)_{mass} + c_w\,K_w}{c\,S + c_w\,S_w}$$

Solving the above equation for S yields

$$S = \frac{c_w}{c} S_w [\frac{K_w}{S_w} - (\frac{K}{S})_{mix}] / [(\frac{K}{S})_{mix} - (\frac{K}{S})_{mass}]$$

If we know everything on the R.H.S. of the above equation then we can solve for S. We do know everything on that side: c and c_w are chosen by us; K_w and S_w we know because we know all about white; and, $(K/S)_{mix}$ and $(K/S)_{mass}$ we can obtain from method one described previously. Once we have S of the colorant we have K from

$$K = S\,(K/S)_{mass}$$

and we are finished.

Method III

The third method is to obtain K and S when we have the colorant mixed with white and then the colorant mixed with black. We thus have two equations:

$$(K/S)_{wm} = \frac{c_w\,k_w + c_{pw}\,k_p}{c_w\,s_w + c_{pw}\,s_p}$$

$$(K/S)_{bm} = \frac{c_b\,k_b + c_{pb}\,k_p}{c_b\,s_b + c_{pb}\,s_p}$$

where c_w, c_b, k_w, s_w, k_b, s_b, c_{pw}, c_{pb}, $(K/S)_{wm}$, $(K/S)_{bm}$ are, repectively: concentration of white, concentration of black, absorption coefficient of white, scattering coefficient of white, absorption coefficient of black, scattering coefficient of black, concentration of colorant (p is for pigment) when mixed with white, concentration of colorant when mixed with black, K/S of white mixture and K/S of black mixture. Solving both equations for k_p and dividing by s_p, we get

$$\frac{k_p}{s_p} = (K/S)_{wm} + \frac{c_w\,s_w}{c_{pw}\,s_p}[(K/S)_{wm} - \frac{k_w}{s_w}]$$

$$\frac{k_p}{s_p} = (K/S)_{bm} + \frac{c_b\,s_b}{c_{pb}\,s_p}[(K/S)_{bm} - \frac{k_b}{s_b}]$$

The L.H.S. of both the above equations are equal. Hence we can equate both R.H.S. and solve for s_p. We get

$$s_p = \frac{c_w\, s_w}{c_{pw}}\left\{\frac{(K/S)_{wm} - \dfrac{k_w}{s_w}}{(K/S)_{bm} - (K/S)_{wm}}\right\} + \frac{c_b\, s_b}{c_{pb}}\left\{\frac{\dfrac{k_b}{s_b} - (K/S)_{bm}}{(K/S)_{bm} - (K/S)_{wm}}\right\}$$

In the above equation we know k_w, s_w, k_b, s_b since we know all about white and black. We know c_w, c_b, c_{pw}, c_{pb} since we choose these values. We can obtain $(K/S)_{wm}$ and $(K/S)_{bm}$ by using the first method of this report. Hence we know everything on the R.H.S. of the above equation and we can calculate s_p. Finally, once we have s_p we have k_p since

$$k_p = (K/S)_{wm}\, s_p + \frac{c_w\, s_w}{c_{pw}}\left[(K/S)_{wm} - \frac{k_w}{s_w}\right]$$

or

$$k_p = (K/S)_{bm}\, s_p + \frac{c_b\, s_b}{c_{pb}}\left[(K/S)_{bm} - \frac{k_b}{s_b}\right]$$

Appendix

We show that as film thickness gets large ($x \rightarrow \infty$) then $\coth(bSx) \rightarrow 1$. Let $u = bSx$. Then

$$\coth(u) = \frac{\cosh(u)}{\sinh(u)}$$

$$= \frac{(e^u + e^{-u})/2}{(e^u - e^{-u})/2}$$

$$= \frac{e^u + \dfrac{1}{e^u}}{e^u - \dfrac{1}{e^u}}$$

$$= \frac{1 + \dfrac{1}{e^{2u}}}{1 - \dfrac{1}{e^{2u}}}$$

$$\rightarrow 1 \text{ as } u \rightarrow \infty$$

datacolor international

Math Specification

Solutions of the Incomplete Hide form of the Kubelka-Munk Equation when K/S > 0 and K/S < 0

August 7, 1995

In a Math Specification document dated Aug. 22, 1990, it is stated that the general solution of the Incomplete Hide form of the Kubelka-Munk equation has two forms; namely

$$R = \frac{1 - R_g\,(a - b\,\coth(bSX))}{a - R_g + b\,\coth(bSX)} \quad \text{when } a^2 > 1$$

and

$$R = \frac{1 - R_g\,(a - b\,\cot(bSX))}{a - R_g + b\,\cot(bSX)} \quad \text{when } a^2 < 1$$

That is to say, they are the same except for the replacement of the hyperbolic cotangent, coth, by the ordinary cotangent, cot, and also the value of b changes from

$$b = (a^2 - 1)^{1/2} \quad \text{when } a^2 > 1$$

to

$$b = (1 - a^2)^{1/2} \quad \text{when } a^2 < 1$$

However, we can think of b as being the same in both equations by the artifice of writing

$$b = (|a^2 - 1|)^{1/2}$$

This definition of b equal to the square root of the absolute value of the square of a minus 1 will take the appropriate form depending on whether the square of a is greater than or less than 1. The other variables in the equation are reflectance, R, background reflectance, R_g , scattering coefficient, S, film thickness, X, and absorption coefficient , K, which shows itself in the variable $a = 1 + K/S$. We note that $K/S > 0 \Rightarrow a^2 > 1$ and $K/S < 0 \Rightarrow a^2 < 1$ (assuming in the latter case that $K/S + 2 > 0$). We shall derive the two forms explicitly.

As a starting point we take the basic equation (16) from page 305 of the chapter "Colorant Formulation and Shading" by Eugene Allen in the book *Optical Radiation Measurements, Volume 2, Color Measurement*, Edited by F. Grum and C.J. Bartleson, which is

$$\int_0^X dx = \int_{R_g}^{R} \frac{d\rho}{S\rho^2 - 2(K+S)\rho + S}$$

We get easily

$$X = \frac{1}{S}\int_{R_g}^{R_s} \frac{d\rho}{\rho^2 - 2\,a\,\rho + 1}$$

The quadratic

$$\rho^2 - 2\,a\,\rho - 1$$

can be written two ways

$$(\rho - a)^2 - (a^2 - 1), \quad \text{when} \quad a^2 > 1$$

or

$$(\rho - a)^2 + (1 - a^2), \quad \text{when} \quad a^2 < 1$$

by completing the square. We note that the second parenthesis in the two forms is positive. Letting

$$u = \rho - a$$

and

$$b = (\,|\,a^2 - 1\,|\,)^{1/2}$$

these two completed squares can be written, respectively, as

$$u^2 - b^2$$

and

$$u^2 + b^2$$

The last written integral equation above then takes the two forms

$$X = \frac{1}{S}\int_{R_g - a}^{R_s - a} \frac{du}{u^2 - b^2} \quad \text{when } a^2 > 1\text{, and } b = (a^2 - 1)^{1/2}$$

and

$$X = \frac{1}{S}\int_{R_g - a}^{R_s - a} \frac{du}{u^2 + b^2} \quad \text{when } a^2 < 1\text{, and } b = (1 - a^2)^{1/2}$$

We shall now deal with each of the above integrals separately.

The following formula

$$\int \frac{du}{u^2 - b^2} = \frac{1}{2b} \ln \frac{u - b}{u + b}$$

is easily verified by taking derivatives.

$$\frac{d}{du}\left(\frac{1}{2b}\ln\frac{u-b}{u+b}\right)=\frac{1}{2b}\frac{\frac{d}{du}\left(\frac{u-b}{u+b}\right)}{\frac{u-b}{u+b}}$$

Since

$$\frac{d}{du}\left(\frac{u-b}{u+b}\right)=\frac{2b}{(u+b)^2}$$

we have

$$\frac{d}{du}\left(\frac{1}{2b}\ln\frac{u-b}{u+b}\right)=\frac{1}{u^2-b^2}$$

Using the above and putting in the limits of integration in the first of the two integrals gives

$$X=\frac{1}{2bS}\left(\ln\frac{u-b}{u+b}\Big|_{R-a}-\ln\frac{u-b}{u+b}\Big|_{R_g-a}\right)$$

or

$$e^{2bSX}=\frac{(R-a-b)(R_g-a+b)}{(R-a+b)(R-a-b)}$$

Solving for R yields

$$R=\frac{(a-b)(R_g-a-b)e^{2bSX}-(a+b)(R_g-a+b)}{(R_g-a-b)e^{2bSX}-(R_g-a+b)}$$

Then using the relations

$$a^2-b^2=1$$

and

$$\coth z=\frac{\cosh z}{\sinh z}=\frac{(e^z+e^{-z})/2}{(e^z-e^{-z})/2}=\frac{e^{2z}+1}{e^{2z}-1}$$

we get the first equation we are looking for

$$R=\frac{1-R_g(a-b\coth(bSX))}{a-R_g+b\coth(bSX)}\qquad a^2>1,\ b=(a^2-1)^{1/2}$$

To evaluate the second integral expression

$$X = \frac{1}{S}\int_{R_g - a}^{R - a} \frac{du}{u^2 + b^2} \qquad a^2 < 1,\ b = (1 - a^2)^{1/2}$$

we use the formula

$$\int \frac{du}{u^2 + b^2} = \frac{1}{b}\tan^{-1}(\frac{u}{b})$$

This is easily verified by using the derivative formula

$$\frac{d}{du}(\tan^{-1} v) = \frac{\frac{dv}{du}}{1 + v^2}$$

Using the integral formula we get

$$bSX = \tan^{-1}\frac{R - a}{b} - \tan^{-1}\frac{R_g - a}{b}$$

Then using the formula

$$\tan(A - B) = \frac{\tan A - \tan B}{1 + \tan A \tan B}$$

we get

$$\tan(bSX) = \frac{b(R - R_g)}{1 + R\,R_g - a(R + R_g)}$$

where we have used , this time ($a^2 < 1$), the relation

$$b^2 = 1 - a^2$$

Taking the reciprocal ($\cot v = 1/\tan v$) and solving for R we get

$$R = \frac{1 - R_g(a - b\cot(bSX))}{a - R_g + b\cot(bSX)} \quad a^2 < 1,\ b = (1 - a^2)^{1/2}$$

and we are finished.

datacolor international

Math Specification

Solution of Incomplete Hide form of the Kubelka-Munk Equation when K or S = 0

August 8, 1995

In a Math Specification document dated August 7, 1995 entitled "Solutions of the Incomplete Hide forms of the Kubelka-Munk Equation when K/S > 0 and K/S < 0", the two following formulas were derived

$$R = \frac{1 - R_g(a - b\,\coth(bSX))}{a - R_g + b\,\coth(bSX)} \quad \text{for } a^2 > 1, \quad b = (a^2 - 1)^{1/2}$$

and

$$R = \frac{1 - R_g(a - b\,\cot(bSX))}{a - R_g + b\,\cot(bSX)} \quad \text{for } a^2 < 1, \quad b = (1 - a^2)^{1/2}$$

The question to be discussed in this document is what happens when $a^2 = 1$. The parameters are as usual R = reflectance, R_g = background reflectance, S = Scattering coefficient, X = film thickness, and a = 1+K/S. From the last formula it is seen that, $a^2 = 1$ implies K/S = 0. In another Math Specification document dated August 22, 1990, the formula for R when K= 0, and S > 0 is given as

$$R = \frac{1 - R_g(1 - \frac{1}{SX})}{1 - R_g + \frac{1}{SX}}$$

It will be shown that this formula for R can be gotten from either the form for R when $a^2 > 1$ or the form when $a^2 < 1$ given above. Besides being of mathematical interest, this is useful since K can converge to zero through positive values or negative (fluorescence) values (and S, of course, is non-negative).

For the first form, we have, when K $\rightarrow$ 0 and thus a $\rightarrow$ 1 and b $\rightarrow$ 0,

$$\coth(bSX) = \frac{e^{2bSX} + 1}{e^{2bSX} - 1}$$

going to infinity since anything not zero raised to the zero power is 1 ($e^0 = 1$) and we have in the limit 1+1 divided by 1-1 or 2 divided by zero which can be considered to be infinite. Hence the equation for R in the limit is

$$R = \frac{1 - R_g(1 - 0*\infty)}{1 - R_g + 0*\infty}$$

The indeterminate form in the above equation results from the product $b*\coth(bSX)$ when $b = 0$. This is dealt with by writing

$$b\coth(bSX) = \frac{b(e^{2bSX}+1)}{e^{2bSX}-1}$$

Letting $b \rightarrow 0$ in both numerator and denominator yields the indeterminate form

$$\frac{0}{0}$$

From Calculus there is available L'Hospital's Rule which says that if

$$\lim_{x\to a}\frac{f(x)}{g(x)} = \frac{0}{0}$$

then

$$\lim_{x\to a}\frac{f(x)}{g(x)} = \lim_{x\to a}\frac{f'(x)}{g'(x)}$$

where the prime indicates the derivative. Applying L'Hospital's Rule yields

$$\lim_{b\to 0}\frac{b(e^{2bSX}+1)}{e^{2bSX}-1} = \lim_{b\to 0}\frac{b(2SXe^{2bSX})+(e^{2bSX}+1)}{2SXe^{2bSX}}$$

The R.H.S. of the above can be simplified to

$$\lim_{b\to 0}(b+\frac{1}{2SX}+\frac{1}{2SXe^{2bSX}})$$

which in the limit is

$$0+\frac{1}{2SX}+\frac{1}{2SX}$$

or

$$\frac{1}{SX}$$

Replacing this for the indeterminate in the above equation for R gives

$$R = \frac{1-R_g(1-\frac{1}{SX})}{1-R_g+\frac{1}{SX}}$$

as desired. Similarly, one gets the same result for the solution when $a^2 < 1$ where the analogous formula

$$\cot(bSX) = \frac{i(e^{2ibSX} + 1)}{e^{2ibSX} - 1}$$

is used. Thus the solution for R is continuous across the boundary $a^2 = 1$ when $K = 0$ and $S > 0$. In the same Math Specification dated August 22, 1990, the result is given for the case when $K = S = 0$; namely,

$$R = R_g$$

This is easily derived by writing the equation for R above when $K = 0$ and $S > 0$ as

$$R = \frac{SX - SXR_g + R_g}{SX - SXR_g + 1}$$

Letting $S = 0$ gives this exact result. The August 22, 1990 document also gives a result for the case when $S = 0$ and K is not zero; namely,

$$R = R_g e^{-2KX}$$

To derive this result, it is necessary to return to the basic equation (16) from page 305 of the chapter "Colorant Formulation and Shading" by Eugene Allen in the book *Optical Radiation Measurements, Volume 2, Color Measurement*, Edited by F. Grum and C.J. Bartleson

$$\int_0^X dx = \int_{R_g}^{R} \frac{d\rho}{S\rho^2 - 2(K+S)\rho + S}$$

Letting $S = 0$, this equation simplifies to

$$X = \int_{R_g}^{R} \frac{d\rho}{-2K\rho}$$

and

$$X = \frac{1}{-2K} \ln(\frac{R}{R_g})$$

which after a little juggling yields the result desired. We note that in the limiting case where $K \rightarrow 0$,

$$R = R_g$$

datacolor international

Programming Note

Summary of Solutions of the Incomplete Hide form of the Kubelka-Munk Equation

August 9, 1995

The solutions of the Kubelka-Munk equation are listed for values of the parameter $a = 1 + K/S$, where K is the Absorption coefficient and S is the Scattering coefficient for the different cases

1) $a^2 > 1$
2) $a^2 < 1$
3) $a^2 = 1$

In the latter case there is a breakdown for the two sub-cases

3.1) $K = 0, S \neq 0$
3.2) $K = 0, S = 0$ $(K/S = 0)$

For completeness, we add the solution for the case where

4) $K \neq 0, S = 0$

For case 1) where $a^2 > 1$, the solution is

$$R = \frac{1 - R_g(a - b\coth(bSX))}{a - R_g + b\coth(bSX)}$$

For case 2) where $a^2 < 1$, the solution is

$$R = \frac{1 - R_g(a - b\cot(bSX))}{a - R_g + b\cot(bSX)}$$

It is to be noted that the only difference between the two above is the use of the hyperbolic cotangent in the first and the use of the ordinary cotangent in the second. These results are derived in a Math Specification document dated August 7, 1995, entitled "Solutions of the Incomplete Hide form of the Kubelka-Munk equation when $K/S > 0$ and $K/S < 0$".

For the $a^2 = 1$ case the solution for the sub-case 3.1) where $K = 0$ and $S > 0$ is

$$R = \frac{1 - R_g(1 - \frac{1}{SX})}{1 - R_g + \frac{1}{SX}}$$

For the sub-case 3.2) $K = 0$, $S = 0$ $(K/S = 0)$, the solution is

$$R = R_g$$

Finally, for the unusual case where $K \neq 0$, $S = 0$, the solution is

$$R = R_g e^{-2KX}$$

The last three solutions are derived in a Math Specification document dated August 8, 1995, entitled Solution of the Incomplete Hide form of the Kubelka-Munk equation when K or $S = 0$.

datacolor international

Math Specification

Overview of Single-Constant Color Matching

August 15, 1995

This report is written for the purpose of giving in greater detail some of the material found in a paper of Denis Martin, dated October 25, 1990, entitled "Textile Recipe Calculations with 1-6 Dyestuffs". Our starting point is a color that we want to match and a collection of n colorants that will be mixed with relative concentrations c_j, j=1,2,...,n, to form a mixture that will yield the closest match possible from this set of colorants to the given color. The given color will be specified by its reflectance curve $R(\lambda)$, where λ is the wavelength in nm, $400 \le \lambda \le 700$ nm. Usually λ is from a discrete set $\{\lambda_i\}$,

$$\lambda_i = 400 + (i-1)*20 \text{ nm}, \quad i=1,2,...,16.$$

Also given is an illuminating light source $S(\lambda)$. Combining the reflectance curve, $R(\lambda)$, the illuminant, $S(\lambda)$, and the eye-sensitivity functions $x(\lambda)$, $y(\lambda)$, $z(\lambda)$, allows us to calculate the tristimulus values X,Y,Z,

$$X(R)=\sum R(\lambda)S(\lambda)x(\lambda)$$

$$Y(R)=\sum R(\lambda)S(\lambda)y(\lambda)$$

$$Z(R)=\sum R(\lambda)S(\lambda)z(\lambda),$$

summations being over the discrete set of λ's. The L, a, b coordinates for X, Y, Z are

$$L = L(Y, Y_n) = L(R)$$

$$a = a(X, Y, X_n, Y_n) = a(R)$$

$$b = b(Y, Z, Y_n, Z_n) = b(R)$$

where X_n, Y_n, Z_n are the tristimulus values for ideal white (the perfect reflecting diffuser) using the same illuminant, S, and observer (x,y,z) combination as used to calculate X,Y,Z. We designate (using a different symbol than D. Martin since his is not available on our system)

$$\Gamma(R) = [L(R), a(R), b(R)].$$

For later use we state the usual color difference formula

$$\Delta E(\Gamma_1, \Gamma_2) = [(\Delta L)^2 + (\Delta a)^2 + (\Delta b)^2]^{1/2}$$

where L_1, a_1, b_1 are the coordinates associated with the reflectance curve R_1 and $\Gamma_1 = \Gamma(R_1)$ and analogously L_2, a_2, b_2 are the coordinates associated with R_2 and $\Gamma_2 = \Gamma(R_2)$, and the differences are:

$$\Delta L = L_1 - L_2 = L(R_1) - L(R_2)$$

$$\Delta a = a_1 - a_2 = a(R_1) - a(R_2)$$

$$\Delta b = b_1 - b_2 = b(R_1) - b(R_2).$$

Now for a set of n colorants with concentrations $\{c_j\}$ there is a concentration vector

$$\underline{c} = (c_1, c_2, \ldots, c_n).$$

Each colorant can be characterized by an absorption coefficient, $k_j(\lambda)$, and a scattering coefficient, $s_j(\lambda)$. In the single-constant theory we have a simplification that results in there being only a single parameter $A_j(\lambda)$ that characterize a colorant instead of two. We let

$$A_j(\lambda) = A_j(\lambda) - A_s(\lambda)$$

where the $A_s(\lambda)$ of the substrate is subtracted off, leaving an $A_j(\lambda)$ that belongs solely to the colorant. Now even though it would seem that the $A_j(\lambda)$ of a colorant should be independent of the concentration, it turns out that this is not true. There is - not yet explainable - a dependency. Hence each individual $A_j(\lambda)$ is a function not only of λ but also c_j, $A_j = A_j(c_j, \lambda)$. Now the typical formula for the $A(\underline{c}, \lambda)$ for the whole mixture is, in the single-constant theory,

$$A(\underline{c}, \lambda) = \sum_{j=1}^{n} c_j\, A_j(c_j, \lambda).$$

Since c_j is a continuous variable with values potentially anywhere from 0 to 1, and since A_j is usually only defined for a discrete grid of c_j values, it is necessary to determine $A_j(c_j, \lambda)$ for all values of c_j between 0 and 1. This can be done in a variety of ways, e.g. using linear interpolation, piecewise linear interpolation, or the use of splines. This will be discussed at another time. We remark at this time, however, that we would do this interpolation for each of the discrete values of λ. Once we have $A(\underline{c}, \lambda)$ then we use the fact that there is some functional relationship in the single-constant theory between A and R,

$$R = R(\lambda) = F(A(\underline{c}, \lambda) = F(A(\lambda)) = F(A).$$

That is to say, for a particular vector of concentrations, $\underline{c}$, defining a mixture of colorants, the resulting mixture has the calculated A, as discussed above, and therefore, by virtue of the functional interconnection between A and R, a reflectance curve, R. For example, in the opaque Kubelka-Munk case the functional relationship is (for all of the discrete λ)

$$R = R(\lambda) = F(A(\lambda)) = 1 + A - [A^2 + 2A]^{1/2}.$$

Now we can state the basic problem under consideration. Consider a color that we want to match using a mixture of colorants at our disposal. The color to be matched has a reflectance curve, R*, which can be obtained from measurements. Now take a starting vector of concentrations $\underline{c}$, presumably a well chosen initial vector; this well chosen starting vector will be discussed another time. Then, as we have discussed above, calculate the A that is associated with this vector. Then find the reflectance, R, associated with this A. From the reflectance's R* and R, we can calculate the tristimulus values

$$(X^*, Y^*, Z^*) = [X(R^*), Y(R^*), Z(R^*)]$$

and

$$(X, Y, Z) = [X(R), Y(R), Z(R)].$$

From these we then calculate the two sets of coordinates

$$\Gamma^* = \Gamma(L^*, a^*, b^*) = (L, a, b)(X^*, Y^*, Z^*)$$

and

$$\Gamma = \Gamma(L, a, b) = (L,a,b)(X,Y,Z).$$

Now we calculate

$$\Delta E[\Gamma^*, \Gamma].$$

Our goal is to make this value as small as possible (hopefully even zero). If the value obtained is not small enough, then we find a new concentration vector

$$\underline{c} + \Delta\, \underline{c}$$

From this we find, successively, a new A, a new R, a new tristimulous vector (X, Y, Z), and finally a new set of L, a, b coordinates. Then we see if our distance measure

$$\Delta E\ [\ \Gamma^*, \Gamma(\underline{c} + \Delta\ \underline{c})]$$

is small enough. If not, we repeat this process, and so on. The basic equation for this process is an analogue of the formula from calculus (Newton's method)

$$f(x + \Delta x) = f(x) + \frac{d(f(x))}{dx}\Delta x$$

which in our vector (multiple variable) case is

$$\Gamma(\underline{c} + \Delta\ \underline{c}\ ) = \Gamma\ (\underline{c}\ ) + J(\frac{L, a, b}{\underline{c}})\Delta\ \underline{c}$$

where J is the Jacobian of the variables L,a,b differentiated with respect to the variables c_j. The formula for the Jacobian is, by successive use of the chain rule for multi-variable function composition,

$$J(\frac{L, a, b}{\underline{c}}) = J(\frac{L, a, b}{X, Y, Z})J(\frac{X, Y, Z}{R})(\frac{dR}{dA})J(\frac{A}{\underline{c}}).$$

Sometimes this method does not converge. In this case an algorithm known as the Levenberg-Marquadt algorithm can be used which involves a combination of Newton's method and the Conjugate Gradient method. This also will be discussed in another report.

We note for reference sake the two-constant theory case differs from the single-constant theory in two ways. First, the variable A as a function of $\underline{c}$ and λ is replaced by K/S where K and S are each individual functions of $\underline{c}$ and λ,

$$K(\underline{c},\lambda) = \sum_{j=1}^{n} c_j k_j(c_j,\lambda)$$

and

$$S(\underline{c},\lambda) = \sum_{j=1}^{n} c_j s_j(c_j,\lambda);$$

and thus

$$(K/S)(\underline{c}\,\lambda) = \frac{K(\underline{c},\lambda)}{S(\underline{c},\lambda)}.$$

The second difference is that the Jacobian formula changes slightly and is in this case

$$J(\frac{L,a,b}{\underline{c}}) = J(\frac{L,a,b}{X,Y,Z})J(\frac{X,Y,Z}{R})J(\frac{R}{K,S})J(\frac{K,S}{\underline{c}}),$$

the difference here is that instead of R being only a function of A, it is now a function of two variables K and S. We note that the derivatives of R with respect to K and S are complicated; they will be discussed separately at another time.

datacolor international

Math Specification

Overview of Single-Constant Color Matching

August 28, 1995

This report is written for the purpose of giving in greater detail some of the material found in a paper of Denis Martin, dated October 25, 1990, entitled "Textile Recipe Calculations with 1-6 Dyestuffs". Our starting point is a color that we want to match and a collection of n colorants that will be mixed with relative concentrations c_j, j=1,2,...,n, to form a mixture that will yield the closest match possible from this set of colorants to the given color. The given color will be specified by its reflectance curve R(λ), where λ is the wavelength in nm, $400 \le \lambda \le 700$ nm. Usually λ is from a discrete set $\{\lambda_i\}$,

$$\lambda_i = 400 + (i-1)*20 \text{ nm}, \quad i=1,2,...,16.$$

Also given is an illuminating light source S(λ). Combining the reflectance curve, R(λ), the illuminant, S(λ), and the eye-sensitivity functions x(λ), y(λ), z(λ), allows us to calculate the tristimulus values X,Y,Z,

$$X = X(R) = \sum R(\lambda)S(\lambda)x(\lambda)$$

$$Y = Y(R) = \sum R(\lambda)S(\lambda)y(\lambda)$$

$$Z = Z(R) = \sum R(\lambda)S(\lambda)z(\lambda),$$

summations being over the discrete set of λ's. The L, a, b coordinates for X, Y, Z are

$$L = L(Y,Y_n) = L(Y(R),Y_n(R)) = L(R)$$

$$a = a(X,Y,X_n,Y_n) = a(X(R),Y(R),X_n(R),Y_n(R)) = a(R)$$

$$b = b(Y,Z,Y_n,Z_n) = b(Y(R),Z(R),Y_n(R),Z_n(R)) = b(R),$$

where X_n, Y_n, Z_n are the tristimulus values for ideal white (the perfect reflecting diffuser) with reflectance $R_{iw}(\lambda)$ and using the same illuminant, $S(\lambda)$, and observer, (x(λ), y(λ), z(λ)), combination as used to calculate X,Y,Z; namely,

$$X_n = X_n(R_{iw}) = \sum R_{iw}(\lambda)S(\lambda)x(\lambda)$$

$$Y_n = Y_n(R_{iw}) = \sum R_{iw}(\lambda)S(\lambda)y(\lambda)$$

$$Z_n = Z_n(R_{iw}) = \sum R_{iw}(\lambda)S(\lambda)z(\lambda).$$

We designate (using a different symbol than D. Martin since his is not available on our system)

$$\Gamma(R) = (L(R), a(R), b(R)).$$

For later use we state the usual color difference formula

$$\Delta E(\Gamma_1, \Gamma_2) = [(\Delta L)^2 + (\Delta a)^2 + (\Delta b)^2]^{1/2},$$

where L_1, a_1, b_1 are the coordinates associated with the reflectance curve R_1 and $\Gamma_1 = \Gamma(R_1)$ and analogously L_2, a_2, b_2 are the coordinates associated with R_2 and $\Gamma_2 = \Gamma(R_2)$, and the differences are:

$$\Delta L = L_1 - L_2 = L(R_1) - L(R_2)$$

$$\Delta a = a_1 - a_2 = a(R_1) - a(R_2)$$

$$\Delta b = b_1 - b_2 = b(R_1) - b(R_2).$$

Now for a set of n colorants with concentrations $\{c_j\}$, there is a concentration vector; namely,

$$\underline{c} = (c_1, c_2, \ldots, c_n).$$

Each of the j-th colorants can be characterized by an absorption coefficient, $k_j(\lambda)$, and a scattering coefficient, $s_j(\lambda)$. In the single-constant theory we have a simplification that results in there being only a single parameter $A_j(\lambda)$ that characterize a colorant instead of two,

$$A_j(\lambda) = (k/s)_j(\lambda), \; j = 1,2,\ldots,n.$$

We let

$$A_j(\lambda) = A_j(\lambda) - A_s(\lambda),$$

where the $A_s(\lambda)$ of the substrate is subtracted off, leaving an $A_j(\lambda)$ that belongs solely to the colorant. Now even though it would seem that the $A_j(\lambda)$ of a colorant should be independent of the concentration, it turns out that this is not true. There is - not yet explainable - a dependency. Hence each individual $A_j(\lambda)$ is a function not only of λ but also c_j, $A_j = A_j(c_j, \lambda)$. Now the typical formula for the $A(\underline{c}, \lambda)$ for the whole mixture is, in the single-constant theory,

$$A(\underline{c}, \lambda) = \sum_{j=1}^{n} c_j \, A_j(c_j, \lambda).$$

Since c_j is a continuous variable with values potentially anywhere from 0 to 1, and since A_j is usually only defined for a discrete grid of c_j values, it is necessary to determine $A_j(c_j, \lambda)$ for all values of c_j between 0 and 1. This can be done in a variety of ways, e.g. using linear interpolation, piecewise linear interpolation, or the use of splines. This interpolation is done for each of the discrete values of λ. Once we have $A(\underline{c}, \lambda)$ then we use the fact that there is some functional relationship in the single-constant theory between A and R,

$$R = R(\lambda) = F(A(\underline{c}, \lambda)) = F(A(\lambda)) = F(A),$$

that is to say, for a particular vector of concentrations, $\underline{c}$, defining a mixture of colorants, the resulting mixture has the calculated A, as discussed above, and therefore, by virtue of the functional interconnection between A and R, a reflectance curve, R. For example, in the opaque Kubelka-Munk case the functional relationship is (for all of the discrete λ)

$$R = R(\lambda) = F(A(\lambda)) = 1 + A - [A^2 + 2A]^{1/2}.$$

Now we can state the basic problem under consideration. Consider a color that we want to match using a mixture of colorants at our disposal. The color to be matched has a reflectance curve, R*, which can be obtained from measurements. Now take a starting vector of concentrations $\underline{c}$, presumably a well chosen initial vector; this well chosen starting vector will be discussed another time. Then, as we have discussed above, calculate the A that is associated with this vector. Then find the reflectance, R, associated with this A. From the reflectance's R* and R, we can calculate the tristimulus values

$$(X^*, Y^*, Z^*) = (X(R^*), Y(R^*), Z(R^*))$$

and

$$(X, Y, Z) = (X(R), Y(R), Z(R)).$$

From these we then calculate the two sets of coordinates

$$\Gamma^* = \Gamma(L^*, a^*, b^*) = (L(Y^*), a(X^*, Y^*), b(Y^*, Z^*))$$

and

$$\Gamma = \Gamma(L, a, b) = (L(Y), a(X,Y), b(Y,Z)),$$

where X_n, Y_n, Z_n are omitted in each of the above since they are the same for both. Now we calculate

$$\Delta E(\Gamma^*, \Gamma).$$

Our goal is to make this value as small as possible (hopefully even zero). If the value obtained is not small enough, then we find a new concentration vector

$$\underline{c} + \Delta\, \underline{c},$$

where in coordinates this vector is

$$(c_1 + \Delta c_1, c_2 + \Delta c_2, \ldots, c_n + \Delta c_n),$$

or

$$\underline{c}' = (c_1', c_2', \ldots, c_n'),$$

with

$$c'_j = c_j + \Delta c_j, \quad j=1,2,\ldots,n$$

From this we find, successively, a new A, where

$$A = A(c'_1, c'_2, \ldots, c'_n) = A(\underline{c}')$$

a new R,

$$R = R(\underline{c}') = F(A(\underline{c}'))$$

a new tristimulous vector, $(X\ Y, Z)$,

$$(X, Y, Z) = (X(R(\underline{c}'), Y(R(\underline{c}'), Z(R(\underline{c}'))),$$

and finally a new set of L, a, b coordinates. Then we see if our distance measure

$$\Delta E\,(\Gamma^*, \Gamma(\underline{c}'))$$

is small enough. If not, we repeat this process, and so on. The basic equation for this process is an analogue of the formula from calculus (Newton's method)

$$f(x+\Delta x) = f(x) + \frac{d(f(x))}{dx}\Delta x,$$

which in our vector (multiple variable) case is

$$\Gamma(\underline{c} + \Delta\,\underline{c}\,) = \Gamma\,(\underline{c}) + J(\frac{L, a, b}{\underline{c}})\Delta\,\underline{c},$$

where J is the Jacobian of the variables L,a,b differentiated with respect to the variables c_j. The formula for the Jacobian is, by successive use of the chain rule for multi-variable function composition,

$$J(\frac{L,a,b}{\underline{c}}) = J(\frac{L,a,b}{X,Y,Z})\,J(\frac{X,Y,Z}{R})\,(\frac{dR}{dA})\,J(\frac{A}{\underline{c}}).$$

Sometimes this method does not converge. In that case an algorithm known as the Levenberg-Marquadt algorithm can be used, which involves a combination of Newton's method and the Conjugate Gradient method.

We note for future reference that for two-constant color matching most of the formulas are the same as in the single-constant case; the differences are given in the following Appendix.

APPENDIX (Two-Constant case)

The two-constant theory case differs from the single-constant theory in two ways. First, the variable A as a function of $\underline{c}$ and λ is replaced by K/S where K and S are each individual functions of $\underline{c}$ and λ,

$$K(\underline{c},\lambda) = \sum_{j=1}^{n} c_j k_j(c_j,\lambda)$$

and

$$S(\underline{c},\lambda) = \sum_{j=1}^{n} c_j s_j(c_j,\lambda);$$

and thus

$$(K/S)(\underline{c}\,\lambda) = \frac{K(\underline{c},\lambda)}{S(\underline{c},\lambda)}.$$

The second difference is that the Jacobian formula changes slightly and is in this case

$$J(\frac{L,a,b}{\underline{c}}) = J(\frac{L,a,b}{X,Y,Z})\, J(\frac{X,Y,Z}{R})\, J(\frac{R}{K,S})\, J(\frac{K,S}{\underline{c}}),$$

the difference here is that instead of R being only a function of A, it is now a function of two variables K and S.

datacolor international

GEOMETRIC MEAN

August 31, 1995

Given two radius vectors in the plane of lengths C_1 and C_2, say $C_1 \le C_2$, from the origin to points (a_1, b_1) and (a_2, b_2) and making angles h_1 and h_2 with the abscissa, respectively. Let

$$\Delta H^2 = 2(C_1C_2 - a_1a_2 - b_1b_2).$$

Extend the shorter of the radii and consider a point on each radius vector that is a distance C, $C_1 \le C \le C_2$, from the origin. Show that if the square of the distance along a line joining the points on both radii is ΔH^2, then

$$C = (C_1C_2)^{1/2},$$

that is to say, C is the geometric mean of C_1 and C_2.

To show this we use the law of cosines and get

$$\Delta H^2 = C^2 + C^2 - 2CC\cos(h_1 - h_2) = 2C^2(1 - \cos(h_1 - h_2)).$$

But

$$\cos(h_1 - h_2) = \cos h_1 \cos h_2 + \sin h_1 \sin h_2.$$

or

$$\cos(h_1 - h_2) = \frac{a_1}{C_1}\frac{a_2}{C_2} + \frac{b_1}{C_1}\frac{b_2}{C_2}.$$

Equating the two expressions for ΔH^2, we get

$$2(C_1C_2 - a_1a_2 - b_1b_2) = 2C^2(1 - \cos(h_1 - h_2)),$$

or, with the substitution for $\cos(h_1 - h_2)$,

$$2(C_1C_2 - a_1a_2 - b_1b_2) = 2C^2\left(\frac{C_1C_2 - a_1a_2 - b_1b_2}{C_1C_2}\right).$$

Hence, after cancellation, we have

$$C^2 = C_1C_2$$

as we wanted to show.

datacolor international

Math Specification

Overview of Single-Constant Color Matching

Sept 7, 1995

This report is written for the purpose of giving in greater detail some of the material found in a paper of Denis Martin, dated October 25, 1990, entitled "Textile Recipe Calculations with 1-6 Dyestuffs". Our starting point is a color that we want to match and a collection of n colorants that will be mixed with relative concentrations c_j, j=1,2,...,n, to form a mixture that will yield the closest match possible from this set of colorants to the given color. The given color will be specified by its reflectance curve $R(\lambda)$, where λ is the wavelength in nm, $400 \le \lambda \le 700$ nm. Usually λ is from a discrete set $\{\lambda_i\}$,

$$\lambda_i = 400 + (i-1)*20 \text{ nm}, \quad i=1,2,\ldots,16.$$

Also given is an illuminating light source $S(\lambda)$. Combining the reflectance curve, $R(\lambda)$, the illuminant, $S(\lambda)$, and the eye-sensitivity functions $x(\lambda)$, $y(\lambda)$, $z(\lambda)$, allows us to calculate the tristimulus values X,Y,Z,

$$X = X(R) = \sum R(\lambda)S(\lambda)x(\lambda)$$

$$Y = Y(R) = \sum R(\lambda)S(\lambda)y(\lambda)$$

$$Z = Z(R) = \sum R(\lambda)S(\lambda)z(\lambda),$$

summations being over the discrete set of λ's. The L, a, b coordinates for X, Y, Z are

$$L = L(Y, Y_n) = L(Y(R), Y_n(R)) = L(R)$$

$$a = a(X, Y, X_n, Y_n) = a(X(R), Y(R), X_n(R), Y_n(R)) = a(R)$$

$$b = b(Y, Z, Y_n, Z_n) = b(Y(R), Z(R), Y_n(R), Z_n(R)) = b(R),$$

where X_n, Y_n, Z_n are the tristimulus values for ideal white (the perfect reflecting diffuser) with reflectance $R_{iw}(\lambda)$ and using the same illuminant, $S(\lambda)$, and observer, $(x(\lambda), y(\lambda), z(\lambda))$, combination as used to calculate X,Y,Z; namely,

$$X_n = X_n(R_{iw}) = \sum R_{iw}(\lambda)S(\lambda)x(\lambda)$$

$$Y_n = Y_n(R_{iw}) = \sum R_{iw}(\lambda)S(\lambda)y(\lambda)$$

$$Z_n = Z_n(R_{iw}) = \sum R_{iw}(\lambda)S(\lambda)z(\lambda).$$

We designate (using a different symbol than D. Martin since his is not available on our word processing system)

$$\underline{\Gamma}(R) = (L(R), a(R), b(R)).$$

For later use we state the usual color difference formula

$$\Delta E(\underline{\Gamma_1}, \underline{\Gamma_2}) = [(\Delta L)^2 + (\Delta a)^2 + (\Delta b)^2]^{1/2},$$

where L_1, a_1, b_1 are the coordinates associated with the reflectance curve R_1 and $\underline{\Gamma_1} = \underline{\Gamma}(R_1)$ and analogously L_2, a_2, b_2 are the coordinates associated with R_2 and $\underline{\Gamma_2} = \underline{\Gamma}(R_2)$, and the differences are:

$$\Delta L = L_1 - L_2 = L(R_1) - L(R_2)$$

$$\Delta a = a_1 - a_2 = a(R_1) - a(R_2)$$

$$\Delta b = b_1 - b_2 = b(R_1) - b(R_2).$$

Now for a set of n colorants with concentrations $\{c_j\}$, there is a concentration vector; namely,

$$\underline{c} = (c_1, c_2, \ldots, c_n)^T.$$

Each of the j-th colorants can be characterized by an absorption coefficient, $k_j(\lambda)$, and a scattering coefficient, $s_j(\lambda)$. In the single-constant theory we have a simplification that results in there being only a single parameter $A_j(\lambda)$ that characterize a colorant instead of two,

$$A_j(\lambda) = (k/s)_j(\lambda),\ j = 1,2,\ldots,n.$$

We let

$$A_j(\lambda) = A_j(\lambda) - A_s(\lambda),$$

where the $A_s(\lambda)$ of the substrate is subtracted off, leaving an $A_j(\lambda)$ that belongs solely to the colorant. Now even though it would seem that the $A_j(\lambda)$ of a colorant should be independent of the concentration, it turns out that this is not true. There is - not yet explainable - a dependency. Hence each individual $A_j(\lambda)$ is a function not only of λ but also c_j, $A_j = A_j(c_j, \lambda)$. Now the typical formula for the $A(\underline{c}, \lambda)$ for the whole mixture is, in the single-constant theory,

$$A(\underline{c}, \lambda) = \sum_{j=1}^{n} c_j\, A_j(c_j, \lambda).$$

Since c_j is a continuous variable with values potentially anywhere from 0 to 1, and since A_j is usually only defined for a discrete grid of c_j values, it is necessary to determine $A_j(c_j, \lambda)$ for all values of c_j

between 0 and 1. This can be done in a variety of ways, e.g. using linear interpolation, piecewise linear interpolation, or the use of splines. This interpolation is done for each of the discrete values of λ. Once we have $A(\underline{c},\lambda)$ then we use the fact that there is some functional relationship in the single-constant theory between A and R,

$$R = R(\lambda) = R(A(\underline{c}, \lambda)) = R(A(\lambda)) = R(A),$$

that is to say, for a particular vector of concentrations, $\underline{c}$, defining a mixture of colorants, the resulting mixture has the calculated A, as discussed above, and therefore, by virtue of the functional interconnection between A and R, a reflectance curve, R. For example, in the opaque Kubelka-Munk case the functional relationship is (for all of the discrete λ)

$$R = R(\lambda) = R(A(\lambda)) = R(A) = 1 + A - [A^2 + 2A]^{1/2}.$$

Now we can state the basic problem under consideration. Consider a color that we want to match using a mixture of colorants at our disposal. The color to be matched has a reflectance curve, R*, which can be obtained from measurements. Now take a starting vector of concentrations $\underline{c}$, presumably a well chosen initial vector; this well chosen starting vector will be discussed another time. Then, as we have discussed above, calculate the A that is associated with this vector. Then find the reflectance, R, associated with this A. From the reflectance's R* and R, we can calculate the tristimulus values

$$(X^*, Y^*, Z^*) = (X(R^*), Y(R^*), Z(R^*))$$

and

$$(X, Y, Z) = (X(R), Y(R), Z(R)).$$

From these we then calculate the two sets of coordinates

$$\underline{\Gamma}^* = \Gamma(L^*, a^*, b^*) = (L(Y^*), a(X^*, Y^*), b(Y^*, Z^*))$$

and

$$\underline{\Gamma} = \underline{\Gamma}(L, a, b) = (L(Y), a(X, Y), b(Y, Z)),$$

where X_n, Y_n, Z_n are omitted in each of the above since they are the same for both. Now we calculate

$$\Delta E(\underline{\Gamma}^*, \underline{\Gamma}).$$

Our goal is to make this value as small as possible (hopefully even zero). If the value obtained is not small enough, then we find a new concentration vector

$$\underline{c} + \Delta\, \underline{c},$$

where in coordinates this vector is

$$(c_1 + \Delta c_1, c_2 + \Delta c_2, \ldots, c_n + \Delta c_n)^T,$$

or

$$\underline{c}' = (c_1', c_2', \ldots, c_n')^T,$$

with

$$c_j' = c_j + \Delta c_j, \quad j=1,2,\ldots,n$$

From this we find, successively, a new A, where

$$A = A(c_1', c_2', \ldots, c_n') = A(\underline{c}')$$

a new R,

$$R = R(A) = R(A(\underline{c}')) = R(\underline{c}')$$

a new tristimulous vector, $(X\ Y, Z)$,

$$(X, Y, Z) = (X(R(\underline{c}'), Y(R(\underline{c}'), Z(R(\underline{c}'))),$$

and finally a new set of L, a, b coordinates. Then we see if our distance measure

$$\Delta E\ (\underline{\Gamma}^*, \underline{\Gamma}(\underline{c}'))$$

is small enough. If not, we repeat this process, and so on. The evaluation of $\underline{\Gamma}(\underline{c}')$ is done approximately. One method that is frequently used is Newton's method which uses the formula from calculus

$$f(x + \Delta x) = f(x) + \frac{d(f(x))}{dx}\Delta x,$$

which yields in our vector (multiple variable) case

$$\underline{\Gamma}(\underline{c} + \Delta\ \underline{c}) = \underline{\Gamma}\ (\underline{c}) + J(\frac{L, a, b}{\underline{c}})\Delta\ \underline{c},$$

or if J is a square matrix (which is the case only when matching with three colorants)

$$\Delta\underline{c} = J^{-1}\ \Delta\underline{\Gamma},$$

where J is the Jacobian of the variables L,a,b differentiated with respect to the variables c_j. The formula for the Jacobian is, by successive use of the chain rule for multi-variable function composition,

$$J(\frac{L,a,b}{\underline{c}}) = J(\frac{L,a,b}{X,Y,Z})\ J(\frac{X,Y,Z,}{R})\ (\frac{dR}{dA})\ J(\frac{A}{\underline{c}}).$$

This method doesn't always converge, especially when the starting vector , $\underline{c}$, is relatively far from the minimum.

Another method used is simply to step in the direction of the negative gradient of $(\Delta E)^2$

$$\Delta \underline{c} = -k\left(\frac{\partial(\Delta E)^2}{\partial c_1}, \frac{\partial(\Delta E)^2}{\partial c_2}, \ldots, \frac{\partial(\Delta E)^2}{\partial c_n}\right)^T = -k\left(\frac{\partial(\Delta E)^2}{\partial \underline{c}}\right),$$

where k is some suitable constant; this method is called the method of Steepest Descent. Since

$$(\Delta E)^2 = (\Delta L)^2 + (\Delta a)^2 + (\Delta b)^2,$$

$$\frac{\partial(\Delta E)^2}{\partial \underline{c}} = -2\left[(L^* - L(\underline{c}))\frac{\partial L}{\partial \underline{c}} + (a^* - a)\frac{\partial a}{\partial \underline{c}} + (b^* - b)\frac{\partial b}{\partial \underline{c}}\right].$$

Thus

$$\frac{\partial(\Delta E)^2}{\partial \underline{c}} \cong -k'\left(\Delta L\frac{\Delta L}{\Delta \underline{c}} + \Delta a\frac{\Delta a}{\Delta \underline{c}} + \Delta b\frac{\Delta b}{\Delta \underline{c}}\right)$$

for k' suitably chosen, or

$$\frac{\partial(\Delta E)^2}{\partial \underline{c}} \cong -k'\left(\frac{(\Delta L)^2 + (\Delta a)^2 + (\Delta b)^2}{\Delta \underline{c}}\right),$$

an expression that will be used later. This method, like Newton's method, also has a problem in that when close to the minimum it can converge in an agonizingly slow fashion.

A method that avoids the pitfalls of both of the above methods is the Levenberg-Marquadt method. It is a combination of both the Steepest Descent method when far from the minimum and switching continuously to the Newton's method when close. The method entails solving for $\Delta \underline{c}$ in the equation

$$(J^T J + \lambda I)\,\Delta \underline{c} = J^T\,\Delta \underline{\Gamma},$$

where J, the Jacobian as specified in Newtons method previously, is a 3 by n matrix, J^T, its transpose, is n by 3, I is the n by n identity matrix, $\Delta \underline{c}$ is n by 1, and $\Delta \underline{\Gamma}$ is 3 by 1. When $\lambda = 0$ the above equation becomes

$$J^T J\,\Delta \underline{c} = J^T \Delta \underline{\Gamma}.$$

Solving this equation for $\Delta \underline{c}$ yields

$$\Delta \underline{c} = (J^T J)^{-1} J^T \Delta \underline{\Gamma}.$$

If J is a square matrix, then

$$\Delta \underline{c} = J^{-1}(J^T)^{-1} J^T \Delta \underline{\Gamma} = J^{-1} I\,\Delta \underline{\Gamma} = J^{-1}\Delta \underline{\Gamma},$$

which is the formula for Newton's method. On the other hand as $\lambda \to \infty$ the basic equation for Levenberg-Marquadt,

$$(J^T J + \lambda I)\Delta \underline{c} = J^T \Delta \underline{\Gamma}$$

becomes diagonally dominant and in the limit approaches

$$\lambda I \, \Delta \underline{c} = J^T \, \Delta \underline{\Gamma} .$$

Solving for $\Delta \underline{c}$ gives

$$\Delta \underline{c} = (\lambda I)^{-1} J^T \Delta \underline{\Gamma} = \frac{1}{\lambda} I^{-1} J^T \Delta \underline{\Gamma} = \frac{1}{\lambda} I \, J^T \Delta \underline{\Gamma} = \frac{1}{\lambda} J^T \Delta \underline{\Gamma} .$$

But

$$J^T = \left(\frac{\partial (L, a, b)}{\partial \underline{c}}\right)^T ,$$

and thus

$$J^T \cong \left(\frac{\Delta (L, a, b)}{\Delta \underline{c}}\right)^T = \left(\left(\frac{\Delta L}{\Delta \underline{c}}\right), \left(\frac{\Delta a}{\Delta \underline{c}}\right), \left(\frac{\Delta b}{\Delta \underline{c}}\right)\right) .$$

Also,

$$\Delta \underline{\Gamma} = \begin{pmatrix} \Delta L \\ \Delta a \\ \Delta b \end{pmatrix} .$$

Hence,

$$J^T \Delta \underline{\Gamma} \cong \left(\frac{\Delta L^2 + \Delta a^2 + \Delta b^2}{\Delta \underline{c}}\right) = \left(\frac{\Delta E^2}{\Delta \underline{c}}\right) .$$

Thus

$$\Delta \underline{c} \cong \frac{1}{\lambda} J^T \Delta \underline{\Gamma} = \frac{1}{\lambda} \left(\frac{\Delta E^2}{\Delta \underline{c}}\right)$$

and

$$\Delta \underline{c} = -k(\frac{\Delta E^2}{\Delta \underline{c}})$$

for some suitable constant k. This is the formula for the method of Steepest Descent.

A convenient starting vector for the iteration process is to use the Least-Squares solution with the following linear approximation for the reflectance,

$$R = 1 + A.$$

We can thus write

$$\Delta L = \frac{\partial L}{\partial Y}\Delta Y$$

$$= \frac{\partial L}{\partial Y}(Y_m - Y_t), \text{ m for match, t for target}$$

$$= \frac{\partial L}{\partial Y}\sum_{\lambda}(R_m - R_t)Sy$$

$$= \frac{\partial L}{\partial Y}\sum_{\lambda}(A_m - A_t)Sy$$

$$= \sum_j c_j \sum_{\lambda} A_j \frac{\partial L}{\partial Y}Sy - \sum A_t \frac{\partial L}{\partial Y}Sy$$

$$= \sum_j c_j F_{1j} - F_{10},$$

where

$$F_{1j} = \sum_{\lambda} A_j \frac{\partial L}{\partial Y}Sy, \text{j=1,2,...,n} \quad \text{and} \quad F_{10} = \sum_{\lambda} A_t \frac{\partial L}{\partial Y}Sy.$$

Similarly,

$$\Delta a = \sum_j c_j F_{2j} - F_{20}$$

where

$$F_{2j} = \sum_{\lambda} A_j(\frac{\partial a}{\partial X}Sx + \frac{\partial a}{\partial Y}Sy), \text{j=1,2,...,n} \quad \text{and} \quad F_{20} = \sum A_t(\frac{\partial a}{\partial X}Sx + \frac{\partial a}{\partial Y}Sy),$$

and

$$\Delta b = \sum_j c_j F_{3j} - F_{30}$$

where

$$F_{3j} = \sum_\lambda A_j(\frac{\partial b}{\partial Y}Sy + \frac{\partial b}{\partial Z}Sz),\ j=1,2,\ldots,n \quad \text{and} \quad F_{30} = \sum A_t(\frac{\partial b}{\partial Y}Sy + \frac{\partial b}{\partial Z}Sz).$$

Since we are in Least-Squares mode, we need to minimize

$$(\Delta L)^2 + (\Delta a)^2 + (\Delta b)^2,$$

that is to say, we need to minimize (multiplying through by $(-1)^2$ for convenience)

$$\sum_{i=1}^{3}(F_{i0} - \sum_{j=1}^{n} c_j F_{ij})^2.$$

In vector matrix form this becomes

$$(\underline{z} - H\underline{c})^T(\underline{z} - H\underline{c}),$$

where

$$\underline{c} = (c_1, c_2, \ldots, c_n)^T$$

$$\underline{z} = (F_{10}, F_{20}, F_{30})^T$$

$$H = \begin{pmatrix} F_{11} & F_{12} & . & . & . & F_{1n} \\ F_{21} & F_{22} & . & . & . & F_{2n} \\ F_{31} & F_{32} & . & . & . & F_{3n} \end{pmatrix}$$

The solution is (see Applied Optimal Estimation edited by Arthur Gelb, M.I.T. Press, 1989)

$$\underline{c} = (H^T H)^{-1} H^T \underline{z}$$

We know from previous work (Howard Witt and Paul Hoffenberg) that this equation carries over into

$$\Delta\underline{c} = (H^T H)^{-1} H^T \underline{v}$$

for suitable $\underline{v}$. Let us compare this formula with Newton's formula

$$\underline{\Gamma}(\underline{c} + \Delta\underline{c}) = \underline{\Gamma}(\underline{c}) + J(\frac{L, a, b}{\underline{c}})\Delta\underline{c}.$$

As indicated previously J is only a square matrix if there are three colorants, so to solve this equation in general for $\Delta \underline{c}$ we use the "pseudoinverse" of J,

$$J^{\#} = (J^T J)^{-1} J^T.$$

We get

$$\Delta \underline{c} = J^{\#} [\, \underline{\Gamma}(\underline{c} + \Delta \underline{c}) - \underline{\Gamma}(\underline{c})] = (J^T J)^{-1} J^T [\Delta \underline{\Gamma}\,].$$

We note the similarity between the two expressions for $\Delta \underline{c}$,

$$\Delta \underline{c} = (H^T H)^{-1} H^T \underline{v},$$

from Least-Squares, and

$$\Delta \underline{c} = (J^T J)^{-1} J^T \Delta \underline{\Gamma}$$

from Newton's method. Let us show that

$$J = H.$$

As we wrote before,

$$J(\frac{L,a,b}{\underline{c}}) = J(\frac{L,a,b}{X,Y,Z}) J(\frac{X,Y,Z}{R})(\frac{dR}{dA}) J(\frac{A}{\underline{c}}).$$

Since

$$A = \sum_j c_j A_j,$$

$$J(\frac{A}{\underline{c}}) = \frac{\partial A}{\partial \underline{c}} = (\frac{\partial A}{\partial c_1}, \frac{\partial A}{\partial c_2}, \ldots, \frac{\partial A}{\partial c_n}) = (A_1, A_2, \ldots, A_n).$$

Since

$$R = 1 + A,$$

$$\frac{dR}{dA} = 1.$$

From

$$X = \sum RSx$$
$$Y = \sum RSy$$
$$Z = \sum RSz$$

we have

$$J(\frac{X,Y,Z}{R}) = \begin{pmatrix} \sum Sx \\ \sum Sy \\ \sum Sz \end{pmatrix}.$$

Finally from

$$L = L(Y)$$
$$a = a(X,Y)$$
$$b = b(Y,Z)$$

we get

$$J(\frac{L,a,b}{X,Y,Z}) = \begin{pmatrix} 0 & \frac{\partial L}{\partial Y} & 0 \\ \frac{\partial a}{\partial X} & \frac{\partial a}{\partial Y} & 0 \\ 0 & \frac{\partial b}{\partial Y} & \frac{\partial b}{\partial Z} \end{pmatrix}.$$

Matrix vector multiplication then shows that

$$J = H.$$

We note for future reference that for two-constant color matching most of the formulas are the same as in the single-constant case; the differences are given in the following Appendix.

APPENDIX (Two-Constant case)

The two-constant theory case differs from the single-constant theory in two ways. First, the variable A as a function of $\underline{c}$ and λ is replaced by K/S where K and S are each individual functions of $\underline{c}$ and λ,

$$K(\underline{c},\lambda) = \sum_{j=1}^{n} c_j k_j(c_j,\lambda)$$

and

$$S(\underline{c},\lambda) = \sum_{j=1}^{n} c_j s_j(c_j,\lambda);$$

and thus

$$(K/S)(\underline{c}\,\lambda) = \frac{K(\underline{c},\lambda)}{S(\underline{c},\lambda)}.$$

The second difference is that the Jacobian formula changes slightly and is in this case

$$J(\frac{L,a,b}{\underline{c}}) = J(\frac{L,a,b}{X,Y,Z})\, J(\frac{X,Y,Z}{R})\, J(\frac{R}{K,S})\, J(\frac{K,S}{\underline{c}}),$$

the difference here is that instead of R being only a function of A, it is now a function of two variables K and S.

datacolor international CONFIDENTIAL

Math Specification

Overview of Single-Constant Color Matching

Sept 11, 1995

This report is written for the purpose of giving in greater detail some of the material found in a paper of Denis Martin, dated October 25, 1990, entitled "Textile Recipe Calculations with 1-6 Dyestuffs". Our starting point is a color that we want to match and a collection of n colorants that will be mixed with relative concentrations c_j, j=1,2,...,n, to form a mixture that will yield the closest match possible from this set of colorants to the given color. The given color will be specified by its reflectance curve R(λ), where λ is the wavelength in nm, $400 \leq \lambda \leq 700$ nm. Usually λ is from a discrete set $\{\lambda_i\}$,

$$\lambda_i = 400 + (i-1)*20 \text{ nm}, \quad i=1,2,...,16.$$

Also given is an illuminating light source S(λ). Combining the reflectance curve, R(λ), the illuminant, S(λ), and the eye-sensitivity functions x(λ), y(λ), z(λ), allows us to calculate the tristimulus values X,Y,Z,

$$X = X(R) = \sum R(\lambda)S(\lambda)x(\lambda)$$

$$Y = Y(R) = \sum R(\lambda)S(\lambda)y(\lambda)$$

$$Z = Z(R) = \sum R(\lambda)S(\lambda)z(\lambda),$$

summations being over the discrete set of λ's. The L, a, b coordinates for X, Y, Z are

$$L = L(Y, Y_m) = L(Y(R), Y_m(R)) = L(R)$$

$$a = a(X, Y, X_m, Y_m) = a(X(R), Y(R), X_m(R), Y_m(R)) = a(R)$$

$$b = b(Y, Z, Y_m, Z_m) = b(Y(R), Z(R), Y_m(R), Z_m(R)) = b(R),$$

where X_m, Y_m, Z_m are the tristimulus values for ideal white (the perfect reflecting diffuser) with reflectance $R_{iw}(\lambda)$ and using the same illuminant, $S(\lambda)$, and observer, (x(λ), y(λ), z(λ)), combination as used to calculate X,Y,Z; namely,

$$X_m = X_m(R_{iw}) = \sum R_{iw}(\lambda)S(\lambda)x(\lambda)$$

$$Y_m = Y_m(R_{iw}) = \sum R_{iw}(\lambda)S(\lambda)y(\lambda)$$

$$Z_m = Z_m(R_{iw}) = \sum R_{iw}(\lambda)S(\lambda)z(\lambda).$$

We designate (using a different symbol than D. Martin since his is not available on our word processing system)

$$\underline{\Gamma}(R) = (L(R), a(R), b(R)).$$

For later use we state the usual color difference formula

$$\Delta E(\underline{\Gamma_1}, \underline{\Gamma_2}) = [(\Delta L)^2 + (\Delta a)^2 + (\Delta b)^2]^{1/2},$$

where L_1, a_1, b_1 are the coordinates associated with the reflectance curve R_1 and $\underline{\Gamma_1} = \underline{\Gamma}(R_1)$ and analogously L_2, a_2, b_2 are the coordinates associated with R_2 and $\underline{\Gamma_2} = \underline{\Gamma}(R_2)$, and the differences are:

$$\Delta L = L_1 - L_2 = L(R_1) - L(R_2)$$

$$\Delta a = a_1 - a_2 = a(R_1) - a(R_2)$$

$$\Delta b = b_1 - b_2 = b(R_1) - b(R_2).$$

Now for a set of n colorants with concentrations $\{c_j\}$, there is a concentration vector; namely,

$$\underline{c} = (c_1, c_2, \ldots, c_n)^T.$$

Each of the j-th colorants can be characterized by an absorption coefficient, $k_j(\lambda)$, and a scattering coefficient, $s_j(\lambda)$. In the single-constant theory we have a simplification that results in there being only a single parameter $A'_j(\lambda)$ that characterize a colorant instead of two,

$$A'_j(\lambda) = (k/s)_j(\lambda),\ j = 1,2,\ldots,n.$$

We let

$$A_j(\lambda) = A'_j(\lambda) - A_s(\lambda),$$

where the $A_s(\lambda)$ of the substrate is subtracted off, leaving an $A_j(\lambda)$ that belongs solely to the colorant. Now even though it would seem that the $A_j(\lambda)$ of a colorant should be independent of the concentration, it turns out that this is not true. There is - not yet explainable - a dependency. Hence each individual $A_j(\lambda)$ is a function not only of λ but also c_j, $A_j = A_j(c_j, \lambda)$. Now the typical formula for the $A(\underline{c}, \lambda)$ for the whole mixture is, in the single-constant theory,

$$A(\underline{c}, \lambda) = \sum_{j=1}^{n} c_j\, A_j(c_j, \lambda).$$

Since c_i is a continuous variable with values potentially anywhere from 0 to 1, and since A_j is usually only defined for a discrete grid of c_j values, it is necessary to determine $A_j(c_j, \lambda)$ for all values of c_j between 0 and 1. This can be done in a variety of ways, e.g. using linear interpolation, piecewise linear interpolation, or the use of splines. This interpolation is done for each of the discrete values of λ. Once we have $A(\underline{c}, \lambda)$ then we use the fact that there is some functional relationship in the single-constant theory between A and R,

$$R = R(\lambda) = R(A(\underline{c}, \lambda)) = R(A(\lambda)) = R(A),$$

that is to say, for a particular vector of concentrations, $\underline{c}$, defining a mixture of colorants, the resulting mixture has the calculated A, as discussed above, and therefore, by virtue of the functional interconnection between A and R, a reflectance curve, R. For example, in the opaque Kubelka-Munk case the functional relationship is (for all of the discrete λ)

$$R = R(\lambda) = R(A(\lambda)) = R(A) = 1 + A - [A^2 + 2A]^{1/2}.$$

Now we can state the basic problem under consideration. Consider a color that we want to match using a mixture of colorants at our disposal. The color to be matched has a reflectance curve, R*, which can be obtained from measurements. Now take a starting vector of concentrations $\underline{c}$, presumably a well chosen initial vector; this well chosen starting vector will be discussed another time. Then, as we have discussed above, calculate the A that is associated with this vector. Then find the reflectance, R, associated with this A. From the reflectance's R* and R, we can calculate the tristimulus values

$$(X^*, Y^*, Z^*) = (X(R^*), Y(R^*), Z(R^*))$$

and

$$(X, Y, Z) = (X(R), Y(R), Z(R)).$$

From these we then calculate the two sets of coordinates

$$\underline{\Gamma}^* = \Gamma(L^*, a^*, b^*) = (L(Y^*), a(X^*, Y^*), b(Y^*, Z^*))$$

and

$$\underline{\Gamma} = \underline{\Gamma}(L, a, b) = (L(Y), a(X, Y), b(Y, Z)),$$

where X_m, Y_m, Z_m are omitted in each of the above since they are the same for both. Now we calculate

$$\Delta E(\underline{\Gamma}^*, \underline{\Gamma}).$$

Our goal is to make this value as small as possible (hopefully even zero). If the value obtained is not small enough, then we find a new concentration vector

$$\underline{c} + \Delta\, \underline{c},$$

where in coordinates this vector is

$$(c_1 + \Delta c_1, c_2 + \Delta c_2, \ldots, c_n + \Delta c_n)^T,$$

or

$$\underline{c}' = (c_1', c_2', \ldots, c_n')^T,$$

with

$$c_j' = c_j + \Delta c_j, \quad j=1,2,\ldots,n$$

From this we find, successively, a new A, where

$$A = A(c_1', c_2', \ldots, c_n') = A(\underline{c}')$$

a new R,

$$R = R(A) = R(A(\underline{c}')) = R(\underline{c}')$$

a new tristimulous vector, $(X\ Y, Z)$,

$$(X, Y, Z)=(X(R(\underline{c}'), Y(R(\underline{c}'), Z(R(\underline{c}'))),$$

and finally a new set of L, a, b coordinates. Then we see if our distance measure

$$\Delta E\ (\underline{\Gamma}^*, \underline{\Gamma}(\underline{c}'))$$

is small enough. If not, we repeat this process, and so on. The evaluation of $\underline{\Gamma}(\underline{c}')$ is done approximately. One method that is frequently used is Newton's method which uses the formula from calculus

$$f(x+\Delta x) = f(x) + \frac{d(f(x))}{dx}\Delta x,$$

which yields in our vector (multiple variable) case

$$\underline{\Gamma}(\underline{c} + \Delta\ \underline{c}) = \underline{\Gamma}\ (\underline{c}) + J(\frac{L, a, b}{\underline{c}})\Delta\ \underline{c},$$

where J is the Jacobian of the variables L,a,b differentiated with respect to the variables c_j. For later use we write this equation as

$$\Delta\underline{\Gamma} = J\ \Delta\underline{c}.$$

In the case that J is a square matrix, we have

$$\Delta\underline{c} = J^{-1}\ \Delta\underline{\Gamma};$$

if J is not square then we can write

$$\Delta\underline{c} = J^{\#}\Delta\underline{\Gamma},$$

where we use the pseudoinverse

$$J^{\#} = (J^TJ)^{-1}J^T,$$

which reduces to the ordinary inverse when J is square.The fact that it is justifiable to call $J^{\#}$ a pseudoinverse follows from the fact that

$$J^{\#}J = [(J^TJ)^{-1}J^T]\,J = (J^TJ)^{-1}(J^TJ) = I$$

where I is the Identity matrix. The formula for the Jacobian is, by successive use of the chain rule for multi-variable function composition,

$$J(\frac{L,a,b}{\underline{c}}) = J(\frac{L,a,b}{X,Y,Z})\,J(\frac{X,Y,Z}{R})\,(\frac{dR}{dA})\,J(\frac{A}{\underline{c}}).$$

We note that Newton's method doesn't always converge (it mignt overshoot), especially when the starting vector , $\underline{c}$, is relatively far from the minimum.

Another method used is the Conjugate Gradient method where

$$\Delta\underline{c} = -k\,(\frac{\partial(\Delta E)^2}{\partial c_1}, \frac{\partial(\Delta E)^2}{\partial c_2}, \ldots, \frac{\partial(\Delta E)^2}{\partial c_n})^T = -k(\frac{\partial(\Delta E)^2}{\partial \underline{c}}),$$

where k is some suitable constant; this method is also called the method of Steepest Descent. Since

$$(\Delta E)^2 = (\Delta L)^2 + (\Delta a)^2 + (\Delta b)^2,$$

$$\frac{\partial(\Delta E)^2}{\partial \underline{c}} = -2\,[(L^* - L(\underline{c}))\frac{\partial L}{\partial \underline{c}} + (a^* - a)\frac{\partial a}{\partial \underline{c}} + (b^* - b)\frac{\partial b}{\partial \underline{c}}].$$

Thus

$$\frac{\partial(\Delta E)^2}{\partial \underline{c}} \cong -k'\,(\Delta L\frac{\Delta L}{\Delta \underline{c}} + \Delta a\frac{\Delta a}{\Delta \underline{c}} + \Delta b\frac{\Delta b}{\Delta \underline{c}}) = -k'(\frac{(\Delta L)^2 + (\Delta a)^2 + (\Delta b)^2}{\Delta \underline{c}}) = -k'(\frac{(\Delta E)^2}{\Delta \underline{c}})$$

for k' suitably chosen, or

$$\frac{(\Delta E)^2}{\Delta \underline{c}} \cong -\frac{1}{-k'}(\frac{\partial(\Delta E)^2}{\partial \underline{c}}),$$

an expression that will be used later. This method, like Newton's method, also has a problem in that when close to the minimum it can converge in an agonizingly slow fashion.

A method that avoids the pitfalls of both of the above methods is the Levenberg-Marquadt method. It is a combination of both the Steepest Descent method when far from the minimum and switching to Newton's method when close. The method entails solving for $\Delta\underline{c}$ in the equation

$$(J^T J + \lambda I)\,\Delta \underline{c} = J^T\,\Delta \underline{\Gamma},$$

where J, the Jacobian as specified in Newtons method previously, is a 3 by n matrix, J^T, its transpose, is n by 3, I is the n by n identity matrix, $\Delta \underline{c}$ is n by 1, and $\Delta \underline{\Gamma}$ is 3 by 1. When $\lambda = 0$ the above equation becomes

$$J^T J\,\Delta \underline{c} = J^T \Delta \underline{\Gamma}.$$

Solving this equation for $\Delta \underline{c}$ yields ($J^T J$ is square and if non-singular can be inverted)

$$\Delta \underline{c} = (J^T J)^{-1} J^T \Delta \underline{\Gamma},$$

or if J is a square matrix, then

$$\Delta \underline{c} = J^{-1}(J^T)^{-1} J^T \Delta \underline{\Gamma} = J^{-1} I\,\Delta \underline{\Gamma} = J^{-1} \Delta \underline{\Gamma},$$

which is the formula for Newton's method. On the other hand as $\lambda \to \infty$ the basic equation for Levenberg-Marquadt,

$$(J^T J + \lambda I)\Delta \underline{c} = J^T \Delta \underline{\Gamma}$$

becomes diagonally dominant and in the limit approaches

$$\lambda I\,\Delta \underline{c} = J^T\,\Delta \underline{\Gamma}.$$

Solving for $\Delta \underline{c}$ gives

$$\Delta \underline{c} = (\lambda I)^{-1} J^T \Delta \underline{\Gamma} = \frac{1}{\lambda} I^{-1} J^T \Delta \underline{\Gamma} = \frac{1}{\lambda} I\, J^T \Delta \underline{\Gamma} = \frac{1}{\lambda} J^T \Delta \underline{\Gamma}.$$

But

$$J^T = \left(\frac{\partial(L,a,b)}{\partial \underline{c}}\right)^T,$$

and thus

$$J^T \cong \left(\frac{\Delta(L,a,b)}{\Delta \underline{c}}\right)^T = \left(\left(\frac{\Delta L}{\Delta \underline{c}}\right), \left(\frac{\Delta a}{\Delta \underline{c}}\right), \left(\frac{\Delta b}{\Delta \underline{c}}\right)\right).$$

Also,

$$\Delta\underline{\Gamma} = \begin{pmatrix} \Delta L \\ \Delta a \\ \Delta b \end{pmatrix}.$$

Hence,

$$J^T \Delta\underline{\Gamma} \cong \left(\frac{\Delta L^2 + \Delta a^2 + \Delta b^2}{\Delta\underline{c}}\right) = \left(\frac{(\Delta E)^2}{\Delta\underline{c}}\right).$$

Thus

$$\Delta\underline{c} \cong \frac{1}{\lambda} J^T \Delta\underline{\Gamma} = \frac{1}{\lambda}\left(\frac{(\Delta E)^2}{\Delta\underline{c}}\right) = \frac{1}{\lambda}\left[\frac{1}{-k'}\left(\frac{\partial(\Delta E)^2}{\partial\underline{c}}\right)\right]$$

and if we let

$$k = \frac{1}{\lambda\, k'}$$

we get

$$\Delta\underline{c} = -k\left(\frac{\partial(\Delta E)^2}{\partial\underline{c}}\right)$$

which is the formula for the method of Steepest Descent.

datacolor international

Math Specification

Least-Squares and Two-Constant Theory

September 14, 1995

We have a target color with reflectance R_t; and, we want to match it with an n-colorant mixture which has reflectance R_m. To do this we shall minimize the sum of the squares of the L,a,b coordinates:

$$(DE)^2 = (DL)^2 + (Da)^2 + (Db)^2 .$$

First, we derive expressins for DL, Da, Db. (In the following, subscript t is used for variables associated with the target and subscript m with variables associated with the matching mixture.

$$DL = L_m - L_t$$

$$\cong \frac{\partial L}{\partial Y} DY$$

$$= \frac{\partial L}{\partial Y} \sum_{\lambda} DR\ E\ y$$, where E is illuminant energy factor and y is the Observer factor

$$\cong \frac{\partial L}{\partial Y} \sum_{\lambda} [\frac{\partial R}{\partial K} DK + \frac{\partial R}{\partial S} DS]\ E\ y ,$$

where $DR \cong \frac{\partial R}{\partial K} DK + \frac{\partial R}{\partial S} DS$

$$= \frac{\partial L}{\partial Y} \sum_{\lambda} [\frac{\partial R}{\partial K}(K_m - K_t) + \frac{\partial R}{\partial S}(S_m - S_t)]\ E\ y$$

$$= \frac{\partial L}{\partial Y} \sum_{\lambda} [\frac{\partial R}{\partial K}(\sum_{i=1}^{n} c_i k_i - K_t) + \frac{\partial R}{\partial S}(\sum_{i=1}^{n} c_i s_i - S_t)]Ey ,$$

where $K_m = \sum_{i=1}^{n} c_i k_i$ and $S_m = \sum_{i=1}^{n} c_i s_i$,

with k_i, s_i absorption and scattering coefficients, respectively, of colorant I

$$= \sum_{i=1}^{n} c_i \sum_{\lambda} [k_i \frac{\partial R}{\partial K} + s_i \frac{\partial R}{\partial S}]\ E\ \frac{\partial L}{\partial Y}\ y$$

$$-\sum_{\lambda}(K_t \frac{\partial R}{\partial K} + S_t \frac{\partial R}{\partial S})\, E \frac{\partial L}{\partial Y}\; y$$

or

$$DL = \sum_{i=1}^{n} c_i F_{i1} - F_{01}$$

$$\text{where } F_{i1} = \sum_{\lambda}[k_i \frac{\partial R}{\partial K} + s_i \frac{\partial R}{\partial S}]\, E \frac{\partial L}{\partial Y}\; y$$

$$\text{and } F_{01} = \sum_{\lambda}[K_t \frac{\partial R}{\partial K} + S_t \frac{\partial R}{\partial S}]\, E \frac{\partial L}{\partial Y}\; y$$

Similarly,

$$Da = \sum_{i=1}^{n} c_i F_{i2} - F_{02}$$

$$\text{where } F_{i2} = \sum_{\lambda}[k_i \frac{\partial R}{\partial K} + s_i \frac{\partial R}{\partial S}]\, E(\frac{\partial a}{\partial X} x + \frac{\partial a}{\partial Y} y),$$

$$\text{and } F_{02} = \sum_{\lambda}[K_t \frac{\partial R}{\partial K} + S_t \frac{\partial R}{\partial S}]\, E(\frac{\partial a}{\partial X} x + \frac{\partial a}{\partial Y} y),$$

x is an Observer factor;

and

$$Db = \sum_{i=1}^{n} c_i F_{i3} - F_{03}$$

$$\text{where } F_{i3} = \sum_{\lambda}[k_i \frac{\partial R}{\partial K} + s_i \frac{\partial R}{\partial S}]\, E(\frac{\partial b}{\partial Y} y + \frac{\partial b}{\partial Z} z),$$

$$\text{and } F_{03} = \sum_{\lambda}[K_t \frac{\partial R}{\partial K} + S_t \frac{\partial R}{\partial S}]\, E(\frac{\partial b}{\partial Y} y + \frac{\partial b}{\partial Z} z),$$

z is an Observer factor.

Hence

$$(DE)^2 = \sum_{j=1}^{3}(\sum_{i=1}^{n} c_i F_{ij} - F_{0j})^2 .$$

A necessary condition for minimization of the above expression is

$$\frac{\partial[(DE)^2]}{\partial c_q} = 0, \quad q = 1,2,\ldots,n.$$

Differentiating yields

$$\frac{\partial[(DE)^2]}{\partial c_q} = \sum_{j=1}^{3} 2\,(\sum_{i=1}^{n} c_i F_{ij} - F_{0j}) F_{qj},\ q = 1,2,\ldots,n\,.$$

Setting these n derivatives to zero gives the n equations,

$$\sum_{i=1}^{n}[\sum_{j=1}^{3} F_{ij} F_{jq}] c_i = \sum_{j=1}^{3} F_{0j} F_{qj},\ q = 1,2,\ldots,n$$

or

$$\sum_{i=1}^{q} a_{iq} c_i = b_{0q},\ q = 1,2,\ldots,n\,,$$

$$\text{where } a_{iq} = \sum_{j=1}^{3} F_{ij} F_{jq} \text{ and } b_{0q} = \sum_{j=1}^{3} F_{0j} F_{qj},\ q = 1,2,\ldots,n\,.$$

In matrix-vector form, the above equation can be written as

$$\mathbf{Ac} = \mathbf{b}$$

where

$$\mathbf{A} = (a_{iq}),\ i = 1,2,\ldots,n,\ q = 1,2,\ldots,n$$

$$\mathbf{c} = (c_1, c_2, \ldots, c_n)^T$$

and

$$\mathbf{b} = (b_1, b_2, \ldots, b_n)^T.$$

Solving for $\mathbf{c}$ gives

$$\mathbf{c} = \mathbf{A}^{-1}\mathbf{b}\,.$$

The formula $\mathbf{c}$ obtained with the above equation should be a good approximation to the formula with the minimum attainable color difference. In other words, if we use the concentration vector $\mathbf{c}$ to synthesize via

$$K_m = \sum_{i=1}^{n} c_i k_i \quad \text{and} \quad S_m = \sum_{i=1}^{n} c_i s_i$$

the reflectance $\mathbf{R}_m$, we don't quite get $\mathbf{R}_m$, but get

$$R_0 = R_m - \delta R\,,$$

where $\mathbf{R}_0$ corresponds to the concentration set $\mathbf{c}$ and $\mathbf{R}_m$ corresponds to the concentration set $\mathbf{c} + \delta\mathbf{c}$.

Part of the reason for not getting the true minimum formula is due to the approximations

$$DL \cong \frac{\partial L}{\partial Y} DY$$

and

$$DR \cong \frac{\partial R}{\partial K} DK + \frac{\partial R}{\partial S} DS.$$

Since the concentration vector **c** yields R_0, not R_m, we can write

$$K_0 = \sum_{i=1}^{n} c_i k_i \quad \text{and} \quad S_0 = \sum_{i=1}^{n} c_i s_i ;$$

and since the concentration $\mathbf{c} + \delta\mathbf{c}$ yields R_m, we can write

$$K_0 + \delta K = \sum_{i=1}^{n} (c_i + \delta c_i) k_i \quad \text{and} \quad S_0 + \delta S = \sum_{i=1}^{n} (c_i + \delta c_i) s_i .$$

Thus by subtraction we get

$$\delta K = \sum_{i=1}^{n} \delta c_i \; k_i \quad \text{and} \quad \delta S = \sum_{i=1}^{n} \delta c_i \; s_i .$$

Now from

$$DR = R_m - R_t = R_0 + \delta R - R_t = R_0 - R_t + \delta R = R_0 - R_t + \frac{\partial R}{\partial K} \delta K + \frac{\partial R}{\partial S} \delta S,$$

the expression

$$DR \cong \frac{\partial R}{\partial K} DK + \frac{\partial R}{\partial S} DS$$

from previously will be replaced in the new circumstance by

$$DR \cong R_0 - R_t + \frac{\partial R}{\partial K} \delta K + \frac{\partial R}{\partial S} \delta S .$$

Hence,

$$DL \cong \frac{\partial L}{\partial Y} DY = \frac{\partial L}{\partial Y} \sum_{\lambda} DR \; E \; y \cong \frac{\partial L}{\partial Y} \sum_{\lambda} [(R_0 - R_t) + \frac{\partial R}{\partial K} \delta K + \frac{\partial R}{\partial S} \delta S] \; E \; y$$

$$= \frac{\partial L}{\partial Y} \sum_{\lambda} [(R_0 - R_t) + \frac{\partial R}{\partial K} \sum_{i=1}^{n} \delta c_i \; k_i + \frac{\partial R}{\partial S} \sum_{i=1}^{n} \delta c_i \; s_i] \, E \, y$$

$$= \sum_{\lambda} (R_0 - R_t) \, E \frac{\partial L}{\partial Y} y + \sum_{i=1}^{n} \delta c_i \sum_{\lambda} (k_i \frac{\partial R}{\partial K} + s_i \frac{\partial R}{\partial S}) \, E \frac{\partial L}{\partial Y} y$$

$$= G_{01} + \sum_{i=1}^{n} F_{i1} \; \delta c_i ,$$

where $G_{01} = \sum_{\lambda} (R_0 - R_t) \, E \frac{\partial L}{\partial Y} y$ and F_{i1} as before, $i = 1,2,\ldots,n$.

Similarly,

$$Da = G_{02} + \sum_{i=1}^{n} \delta c_i F_{i2}$$

where $G_{02} = \sum_{\lambda} (R_0 - R_t) \, E \, (\frac{\partial a}{\partial X} x + \frac{\partial a}{\partial Y} y)$ and F_{i2} as before,

$i = 1,2,\ldots,n$;

and

$$Db = G_{03} + \sum_{i=1}^{n} \delta c_i \; F_{i3}$$

where $G_{03} = \sum_{\lambda} (R_0 - R_t) \, E \, (\frac{\partial b}{\partial Y} y + \frac{\partial b}{\partial Z} z)$ and F_{i3} as before,

$i = 1,2,\ldots,n$.

Hence,

$$(DE)^2 = \sum_{j=1}^{3} (G_{0j} + \sum_{i=1}^{n} \delta c_i \; F_{ij})^2 .$$

As before a necessary condition for minimization is that the partials are equal to zero. The partials are:

$$\frac{\partial [(DE)]^2}{\partial \, \delta c_q} = \sum_{j=1}^{3} 2(G_{0j} + \sum_{i=1}^{n} \delta c_i \; F_{ij}) \, F_{qj} , \quad q = 1,2,\ldots,n.$$

Setting these equal to zero yields

$$\sum_{i=1}^{n} (\sum_{j=1}^{3} F_{ij} \; F_{qj}) \, \delta c_i = - \sum_{j=1}^{3} G_{0j} F_{qj} , \quad q = 1,2,\ldots,n$$

or

$$\sum_{i=1}^{n} a_{iq}\, \delta c_i = d_{0i}, \quad \text{where } a_{iq} = \sum_{j=1}^{3} F_{ij}\, F_{qj} \text{ and } d_{0q} = \sum_{j=1}^{3} G_{0j}\, F_{qj},$$

$$q = 1,2,\ldots,n.$$

In matrix vetor form the above can be written as

$$\mathbf{A}\, \delta\mathbf{c} = \mathbf{d}$$

where $\mathbf{A}$ is as before,

$$\delta\mathbf{c} = (\delta c_1, \delta c_2, \ldots, \delta c_n)^T,$$

and

$$\mathbf{d} = (d_1, d_2, \ldots, d_n)^T.$$

Therefore,

$$\delta\mathbf{c} = \mathbf{A}^{-1}\mathbf{d}.$$

We note that the same inverted matrix that is used, in the above equation, to calculate the corrections is used to calculate, as shown previously, the approximate formula:

$$\mathbf{c} = \mathbf{A}^{-1}\mathbf{b}.$$

The significance of this is that the matrix used to get our first approximation can then be used again and again to converge as close as we wish to the optimal match.

datacolor international

Math Specification

Textile Recipe Calculation

September 22, 1995

Introduction

This document will give more details on some of the material in the document "Textile Recipe Calculations with 1 to 6 Dyestuffs", by Denis Martin, dated October 25, 1990. It describes the method used for recipe calculations and correction for textile coloring.

Problem Definition

Given a color reflectance curve defined by the function $R(\lambda)$, where λ is the wavelength and has domain [380, 700] nanometers (nm). In practice we use only a discrete subset of the domain, usually $\{\lambda_j\}$, j=1,2,...,16, where

$$\lambda_j = 380 + j*20, \quad j = 1,2,\ldots,16.$$

The goal of recipe calculation is to find a subset $\underline{D}$ of given dyestuffs

$$\underline{D} = \{D_1, D_2, \ldots, D_n\}, \; n \le 6$$

where D_i is dyestuff i with a given characteristic.

Any dyestuff D_i has associated with two parameters $k_i = k_i(\lambda)$, the absorption coefficient, and $s_i = s_i(\lambda)$, the scattering (diffusion) coefficient. Its reflection $R_i(\lambda)$ is related to k_i and s_i by the famous Kubelka-Munk formula (opaque case)

$$\frac{k_i}{s_i}(\lambda) = \frac{k_i(\lambda)}{s_i(\lambda)} = \frac{(1 - R_i(\lambda))^2}{2\,R_i(\lambda)}$$

or its inverse

$$R_i = 1 + \frac{k_i(\lambda)}{s_i(\lambda)} - \left[\left(\frac{k_i(\lambda)}{s_i(\lambda)}\right)^2 + 2\left(\frac{k_i(\lambda)}{s_i(\lambda)}\right)\right]^{\frac{1}{2}}.$$

In the textile domain the functions $s_i(\lambda)$ are often constant , s^*, and we use the letter A_i' for the absorption of the dyestuff D_i and identify it with the relation

$$A_i'(\lambda) = \frac{k_i(\lambda)}{s_i(\lambda)} = \frac{k_i(\lambda)}{s^*},$$

and then define

$$A_i(\lambda) = A_i'(\lambda) - A_{substrate}.$$

The combination $\underline{D}$ of many dyestuffs with concentrations $\mathbf{c}$ where

$$\mathbf{c} = (c_1, c_2, \ldots, c_n)$$

on a given substrate results in an absorption $A(\mathbf{c})$ which is to first approximation

$$A(\mathbf{c}) = A(\mathbf{c}, \lambda) = \sum_{i=1}^{n} c_i A_i(c_i, \lambda).$$

For a given reflectance curve $R(\lambda)$ under illuminant $I(\lambda)$ with eye-sensitivity function $(x_{10}(\lambda), y_{10}(\lambda), z_{10}(\lambda))$, we have three color coordinates (X,Y,Z) given by the formulas

$$X = \sum_{\lambda=380}^{700} R(\lambda) I(\lambda) x_{10}(\lambda)$$

$$Y = \sum_{\lambda=380}^{700} R(\lambda) I(\lambda) y_{10}(\lambda)$$

$$Z = \sum_{\lambda=380}^{700} R(\lambda) I(\lambda) z_{10}(\lambda).$$

Usually we work with L,a,b coordinates.

$$L = L(Y) = L(Y(R)) = L(R)$$

$$a = a(X,Y) = a(X(R), Y(R)) = a(R)$$

$$b = b(Y,Z) = b(Y(R), Z(R) = b(R).$$

For later use we note that the difference of L,a,b coordinates between two reflectances R and r is defined as:

$$\mathbf{dE}(R, r, I) = [L(R) - L(r), a(R) - a(r), b(R) - b(r)]$$

for some illuminant I. Now we can state our problem: For a given reflectance $R(\lambda)$ find a subset of dyestuffs

$$\underline{D} = (D_1, D_2, \ldots, D_n)$$

where dyestuff D_i has absorption A_i, i=1,2,...,n and a concentration vector

$$\mathbf{c} = (c_1, c_2, \ldots, c_n)$$

such that

$$R = 1 + A - [A^2 + 2A]^{\frac{1}{2}},$$

where

$$A(\mathbf{c}) = \sum_{i=1}^{n} c_i A(c_i).$$

Characterization of a Dyestuff

A_i is a function not only of the wavelength λ, but of the concentration c_i of dyestuff D_i, i=1,2,...,n; it represents the part of "the energy" which remains on the substrate at a given wavelength λ. As a function of c_i, $A_i(c_i)$ is usually positive and montone increasing. The problem that has to be dealt with is that $A_i(c_i)$ is initially only defined for a discrete set of values $\{c_i^k\}$ for k=1,2,...,m and therefore what do we do when we need $A_i(c_i)$ for a value of c_i which is not one of the $\{c_i^k, k = 1,2,\ldots,m\}$. In that case we interpolate, using a piecewise linear function, polynomial, or by means of spline representation. Splines seem to be the favored method today. At each node c_i^j, we know an ordinate $A_i^j = A_i(c_i^j)$. Each point c in $[0, c_i^m]$ belongs to a subinterval $[c_i^j, c_i^{j+1}]$; we can write for this c the following formula

$$A_i(c) = d(c)A_i^j + e(c)A_i^{j+1} + f(c)\left(\frac{dA_i^j(c_i^j)}{dc_i}\right) + g(c)\left(\frac{dA_i^{j+1}(c_i^{j+1})}{dc_i}\right).$$

The four functions d(c), e(c), f(c), g(c) are polynomials of degree 3.

Solution of Colometric Equation for a given reflectance R*(λ)

Given a reflectance R*(λ). From it we can calculate tristimulus values X*, Y*, Z*,

$$X^* = X(R^*)$$
$$Y^* = Y(R^*)$$
$$Z^* = Z(R^*)$$

From the latter we can calculate L,a,b coordinates L*,a*,b*

$$L^* = L(Y^*)$$
$$a^* = a(X^*, Y^*)$$
$$b^* = b(Y^*, Z^*)$$

Our problem is to find a subset of dyestuffs $\underline{D} = \{D_1, D_2, \ldots, D_n)$ with associated absorptions $\{A_1, A_2, \ldots, A_n\}$ and a vector $\mathbf{c} = (c_1, c_2, \ldots, c_n)$ such that for

$$A(\mathbf{c}) = \sum_{i=1}^{n} c_i A_i(c_i),$$

$$R(\mathbf{c}) = R(A(\mathbf{c})),$$

$$X(\mathbf{c}) = X(R(\mathbf{c}))$$
$$Y(\mathbf{c}) = Y(R(\mathbf{c}))$$
$$Z(\mathbf{c}) = Z(R(\mathbf{c}))$$

and

$$L = L(\mathbf{c}) = L(Y(\mathbf{c}))$$
$$a = a(\mathbf{c}) = a(X(\mathbf{c}), Y(\mathbf{c}))$$
$$b = b(\mathbf{c}) = b(Y(\mathbf{c}), Z(\mathbf{c}))$$

then is

$$(L, a, b) \cong (L^*, a^*, b^*),$$

that is to say, is

$$(L - L^*)^2 + (a - a^*)^2 + (b - b^*)^2$$

sufficiently small. If this is not the case, then find a small perturbation of $\mathbf{c}$, namely $\mathbf{dc}$, such that if we calculate using Taylor's formula

$$(L(\mathbf{c} + \mathbf{dc}), a(\mathbf{c} + \mathbf{dc}), b(\mathbf{c} + \mathbf{dc}))^T \cong (L(\mathbf{c}), a(\mathbf{c}), b(\mathbf{c}))^T + \mathbf{J}(L(\mathbf{c}), a(\mathbf{c}), b(\mathbf{c}))\mathbf{dc}$$

then we see if

$$(L(\mathbf{c} + \mathbf{dc}), a(\mathbf{c} + \mathbf{dc}), b(\mathbf{c} + \mathbf{dc})) \cong (L^*, a^*, b^*).$$

If such is the case we are finished; otherwise we repeat this process, and so on.

Metamerism

We consider now 2 illuminants I_1, I_2, and two reflectance curves R and r. We use the notation

(L_1^R, a_1^R, b_1^R); (L_2^R, a_2^R, b_2^R); (L_1^r, a_1^r, b_1^r); (L_2^r, a_2^r, b_2^r) for L,a,b coordinates in the following four combinations, respectively: illuminant I_1, reflectance R; illuminant I_2, reflectance R; illuminant I_1, reflectance r; illuminant I_2, reflectance r. The metamerism is defined as follows:

$$\mathbf{dM}(R, r, I_1, I_2) = \left(L_1^R \frac{L_2^r}{L_1^r} - L_2^R,\ a_1^R \frac{a_2^r}{a_1^r} - a_2^R,\ b_1^R \frac{b_2^r}{b_1^r} - b_2^R \right).$$

Let us look at the first of the coordinates of the above 3 vector. First we multiply by L_1^r which gives us

$$L_1^R L_2^r - L_2^R L_1^r.$$

Then we add and subtract $L_1^R L_2^R$ getting

$$L_1^R L_2^r - L_2^R L_1^r + L_1^R L_2^R - L_1^R L_2^R.$$

Rearranging, this can be written

$$L_2^R (L_1^R - L_1^r) - L_1^R (L_2^R - L_2^r)$$

Similarly, working with the second and third coordinates of $\mathbf{dM}$, we get them equal to, respectively,

$$a_2^R (a_1^R - a_1^r) - a_1^R (a_2^R - a_2^r)$$

and

$$b_2^R (b_1^R - b_1^r) - b_1^R (b_2^R - b_2^r).$$

Now from

$$\mathbf{dE}(R, r, I_1) = [L_1^R - L_1^r,\ a_1^R - a_1^r,\ b_1^R - b_1^r]$$

and

$$\mathbf{dE}(R, r, I_2) = [L_2^R - L_2^r,\ a_2^R - a_2^r,\ b_2^R - b_2^r],$$

if

$$\mathbf{dE}(R, r, I_1) = \mathbf{0} \text{ and } \mathbf{dE}(R, r, I_2) = \mathbf{0},$$

then all the six expressions enclosed in parentheses in the above three expressions for the coordinates of $\mathbf{dM}$ are equal to zero; this implies $\mathbf{dM} = \mathbf{0}$. On the other hand if

$$\mathbf{dM}(R, r, I_1, I_2) = \mathbf{0} \text{ and one of } \mathbf{dE}(R, r, I) = \mathbf{0},$$

then a little examination of the above three expressions show that both

$$\mathbf{dE}(R, r, I_1) = \mathbf{0} \text{ and } \mathbf{dE}(R, r, I_2) = \mathbf{0}.$$

To show this, let us assume

$$\mathbf{dE}(R, r, I_1) = [L_1^R - L_1^r, a_1^R - a_1^r, b_1^R - b_1^r] = \mathbf{0}.$$

Then

$$L_1^R - L_1^r = 0$$
$$a_1^R - a_1^r = 0$$
$$b_1^R - b_1^r = 0$$

From (We only work with the first coordinate since everything is similar for coordinates two and three.)

$$\mathbf{dM} = \left(\frac{1}{L_1^r}(L_1^R L_2^r - L_2^R L_1^r), \ldots, \ldots \right) = \mathbf{0}$$

and again adding and subtracting the same thing $L_1^R L_2^R$ and then rearranging as before, we get

$$\mathbf{dM} = \left(\frac{1}{L_1^r}[L_2^R(L_1^R - L_1^r) - L_1^R(L_2^R - L_2^r)], \ldots, \ldots \right) = \mathbf{0}.$$

Thus since each coordinate of the above vector must be zero, and because of our assumption that $\mathbf{dE}(R, r, I_1) = \mathbf{0}$ which means

$$L_1^R - L_1^r = 0,$$

we have

$$L_2^R - L_2^r = 0.$$

Similarly, we find that

$$a_2^R - a_2^r = 0$$

and

$$b_2^R - b_2^r = 0.$$

This shows that it is the same to solve two standard problems $\mathbf{dE}(R, r, I_i) = \mathbf{0}, i = 1,2$ or to solve the $\mathbf{dE}$ equation for the first illuminant and the $\mathbf{dM} = \mathbf{0}$ for both illuminants.

$$\Delta\underline{\Gamma} = \begin{pmatrix} \Delta L \\ \Delta a \\ \Delta b \end{pmatrix}.$$

Hence,

$$J^T \Delta\underline{\Gamma} \cong \left(\frac{\Delta L^2 + \Delta a^2 + \Delta b^2}{\Delta\underline{c}}\right) = \left(\frac{(\Delta E)^2}{\Delta\underline{c}}\right).$$

Thus

$$\Delta\underline{c} \cong \frac{1}{\lambda} J^T \Delta\underline{\Gamma} \doteq \frac{1}{\lambda}\left(\frac{(\Delta E)^2}{\Delta\underline{c}}\right) = \frac{1}{\lambda}\left[\frac{1}{-k'}\left(\frac{\partial(\Delta E)^2}{\partial\underline{c}}\right)\right]$$

and if we let

$$k = \frac{1}{\lambda\, k'}$$

we get

$$\Delta\underline{c} = -k\left(\frac{\partial(\Delta E)^2}{\partial\underline{c}}\right)$$

which is the formula for the method of Steepest Descent.

APPENDIX I

A convenient starting vector for the iteration process is to use the Least-Squares solution with the following linear approximation for the reflectance,

$$R = 1 + A .$$

We can thus write

$$\Delta L = \frac{\partial L}{\partial Y}\Delta Y$$

$$= \frac{\partial L}{\partial Y}(Y_m - Y_t), \text{ m for match, t for target}$$

$$= \frac{\partial L}{\partial Y}\sum_{\lambda}(R_m - R_t)Sy$$

$$= \frac{\partial L}{\partial Y}\sum_{\lambda}(A_m - A_t)Sy$$

$$= \sum_j c_j \sum_{\lambda} A_j \frac{\partial L}{\partial Y}Sy - \sum A_t \frac{\partial L}{\partial Y}Sy$$

$$= \sum_j c_j F_{1j} - F_{10} ,$$

where

$$F_{1j} = \sum_{\lambda} A_j \frac{\partial L}{\partial Y}Sy , j=1,2,...,n \quad \text{and} \quad F_{10} = \sum_{\lambda} A_t \frac{\partial L}{\partial Y}Sy .$$

Similarly,

$$\Delta a = \sum_j c_j F_{2j} - F_{20}$$

where

$$F_{2j} = \sum_{\lambda} A_j (\frac{\partial a}{\partial X}Sx + \frac{\partial a}{\partial Y}Sy) , j=1,2,...,n \quad \text{and} \quad F_{20} = \sum A_t (\frac{\partial a}{\partial X}Sx + \frac{\partial a}{\partial Y}Sy),$$

and

$$\Delta b = \sum_j c_j F_{3j} - F_{30}$$

where

$$F_{3j} = \sum_{\lambda} A_j(\frac{\partial b}{\partial Y} Sy + \frac{\partial b}{\partial Z} Sz),\ j=1,2,\ldots,n \quad \text{and} \quad F_{30} = \sum A_t(\frac{\partial b}{\partial Y} Sy + \frac{\partial b}{\partial Z} Sz).$$

Since we are in Least-Squares mode, we need to minimize

$$(\Delta L)^2 + (\Delta a)^2 + (\Delta b)^2,$$

that is to say, we need to minimize (multiplying through by $(-1)^2$ for convenience)

$$\sum_{i=1}^{3} (F_{i0} - \sum_{j=1}^{n} c_j F_{ij})^2.$$

In vector matrix form this becomes

$$(\underline{z} - H\underline{c})^T (\underline{z} - H\underline{c}),$$

where

$$\underline{c} = (c_1, c_2, \ldots, c_n)^T$$

$$\underline{z} = (F_{10}, F_{20}, F_{30})^T$$

$$H = \begin{pmatrix} F_{11} & F_{12} & . & . & . & F_{1n} \\ F_{21} & F_{22} & . & . & . & F_{2n} \\ F_{31} & F_{32} & . & . & . & F_{3n} \end{pmatrix}$$

The solution is (see Applied Optimal Estimation edited by Arthur Gelb, M.I.T. Press, 1989)

$$\underline{c} = (H^T H)^{-1} H^T \underline{z}$$

We know from previous work (Howard Witt and Paul Hoffenberg) that this equation carries over into

$$\Delta\underline{c} = (H^T H)^{-1} H^T \underline{v}$$

for suitable $\underline{v}$. Let us compare this formula with Newton's formula

$$\underline{\Gamma}(\underline{c} + \Delta\underline{c}) = \underline{\Gamma}(\underline{c}) + J(\frac{L, a, b}{\underline{c}})\Delta\underline{c}.$$

As indicated previously we can solve this equation in general for $\Delta\underline{c}$ using the "pseudoinverse" of J,

$$J^{\#} = (J^T J)^{-1} J^T,$$

and obtain

$$\Delta \underline{c} = J^{\#} \, [\, \underline{\Gamma}(\underline{c} + \Delta \underline{c}) - \underline{\Gamma}(\underline{c})] = (J^T J)^{-1} J^T [\Delta \underline{\Gamma}\,].$$

We note the similarity between this expressions for $\Delta \underline{c}$, and the one from Least-Squares

$$\Delta \underline{c} = (H^T H)^{-1} H^T \underline{v},$$

for some suitable vector $\underline{v}$. Without considering what the vector $\underline{v}$ must be ($\Delta \underline{\Gamma}$ and why) let us show that

$$J = H.$$

As we wrote before,

$$J(\frac{L,a,b}{\underline{c}}) = J(\frac{L,a,b}{X,Y,Z}) J(\frac{X,Y,Z}{R})(\frac{dR}{dA}) J(\frac{A}{\underline{c}}).$$

Since

$$A = \sum_j c_j A_j,$$

$$J(\frac{A}{\underline{c}}) = \frac{\partial A}{\partial \underline{c}} = (\frac{\partial A}{\partial c_1}, \frac{\partial A}{\partial c_2}, \ldots, \frac{\partial A}{\partial c_n}) = (A_1, A_2, \ldots, A_n).$$

Since

$$R = 1 + A,$$

$$\frac{dR}{dA} = 1.$$

From

$$X = \sum RSx$$
$$Y = \sum RSy$$
$$Z = \sum RSz$$

we have

$$J\left(\frac{X,Y,Z}{R}\right) = \begin{pmatrix} \sum Sx \\ \sum Sy \\ \sum Sz \end{pmatrix}.$$

Finally from

$$\begin{aligned} L &= L(Y) \\ a &= a(X,Y) \\ b &= b(Y,Z) \end{aligned}$$

we get

$$J\left(\frac{L,a,b}{X,Y,Z}\right) = \begin{pmatrix} 0 & \frac{\partial L}{\partial Y} & 0 \\ \frac{\partial a}{\partial X} & \frac{\partial a}{\partial Y} & 0 \\ 0 & \frac{\partial b}{\partial Y} & \frac{\partial b}{\partial Z} \end{pmatrix}.$$

Matrix vector multiplication then shows that

$$J = H.$$

APPENDIX II

The two-constant theory case differs from the single-constant theory in two ways. First, the variable A as a function of $\underline{c}$ and λ is replaced by K/S where K and S are each individual functions of $\underline{c}$ and λ,

$$K(\underline{c},\lambda) = \sum_{j=1}^{n} c_j k_j(c_j,\lambda)$$

and

$$S(\underline{c},\lambda) = \sum_{j=1}^{n} c_j s_j(c_j,\lambda);$$

and thus

$$(K/S)(\underline{c}\,\lambda) = \frac{K(\underline{c},\lambda)}{S(\underline{c},\lambda)}.$$

The second difference is that the Jacobian formula changes slightly and is in this case

$$J(\frac{L,a,b}{\underline{c}}) = J(\frac{L,a,b}{X,Y,Z})\, J(\frac{X,Y,Z}{R})\, J(\frac{R}{K,S})\, J(\frac{K,S}{\underline{c}}),$$

the difference here is that instead of R being only a function of A, it is now a function of two variables K and S.

The Kalman Filter and Iterated Least Squares

The Kalman Filter
and
Iterated Least - Squares

For the system model

$$x_k = \phi_{k-1} x_{k-1} + W_{k-1}, \quad W_k \sim N(0, Q_k)$$

and measurement model

$$z_k = H_k x_k + V_k, \quad V_k \sim N(0, R_k)$$

with initial conditions

$$E[X(0)] = \hat{X}_0$$

$$E[(X(0) - \hat{X}_0)(X(0) - \hat{X}_0)^T] = P_0$$

and the assumption that

$$E[w_k V_j^T] = 0 \quad \text{for all } j, k$$

we have the following five equations for the discrete Kalman Filter.

(2)

1) $x_k(-) = \phi_{k-1}\,\hat{x}_{k-1}(+)$

2) $P_k(-) = \phi_{k-1}\,P_{k-1}(+)\,\phi_{k-1}^T + Q_{k-1}$

3) $\hat{X}_k(+) = \hat{X}_k(-) + K_k\left[z_k - H_k\,\hat{X}_k(-)\right]$

4) $P_k(+) = \left[I - K_k\,H_k\right]P_k(-)$

5) $K_k = P_k(-)\,H_k^T\left[H_k\,P_k(-)\,H_k^T + R_k\right]^{-1}$

Equations 1) and 2) are, respectively, the state estimate and error covariance extrapolations between measurements.

Equations 3) and 4) are, respectively, the discontinuous state estimate and error covariance updates across a measurement. Equation 5) is the formula for the Kalman Gain matrix.

(3)

Equations 3) and 4) in words can be looked at in the following way. Given an estimate $\hat{X}_k(-)$ and the error covariance matrix $P_k(-)$, if we get another measurement value $z_k = H_k x_k + v_k$, then we can calculate K_k from equation 5) and obtain equation 3) for $\hat{X}_k(+)$

6) $$\hat{X}_k(+) = \hat{X}_k(-) + K_k\left[z_k - H_k \hat{X}_k(-)\right]$$

What we plan to do is to find an alternative formula for K_k, namely,

7) $$K_k = P_k(+) H_k^T R_k^{-1}$$

and thus be able to write 3) as

8) $$\hat{X}_k(+) = \hat{X}_k(-) + P_k(+) H_k^T R_k^{-1}\left[z_k - H_k \hat{X}_k(-)\right]$$

and then develop this exact same equation with suitable definitions in the iterated least-squares formulation.

④

In order to derive 7) we first show that

9) $$P_k^{-1}(+) = P_k^{-1}(-) + H_k^T R_k^{-1} H_k$$

We do this by showing that from equation 4) for $P_k(+)$ and 9) for $P_k^{-1}(+)$ we have, as one would expect,

10) $$P_k(+) P_k^{-1}(+) = I$$

Dropping subscripts we write

$$P(+) P^{-1}(+) = \{[I-KH]P(-)\}\{P^{-1}(-) + H^T R^{-1} H\}$$

$$= [I-KH][I + P(-) H^T R^{-1} H]$$

$$= I - KH + P(-)H^T R^{-1} H - KHP(-)H^T R^{-1} H$$

Substituting from equation 5) for K, where again we have dropped subscripts, we can continue the chain of equalities as

⑤

$$P(+)P^{-1}(+) = I - \{P(-)H^T[HP(-)H^T + R]^{-1}\}H$$

$$+P(-)H^TR^{-1}H$$

$$-\{P(-)H^T[HP(-)H^T + R]^{-1}HP(-)H^TR^{-1}H$$

$$= I - P(-)H^T\{[HP(-)H^T + R]^{-1} - R^{-1}\}H$$

$$-P(-)H^T\{[HP(-)H^T + R]^{-1}HP(-)H^TR^{-1}\}H$$

$$= I - P(-)H^T\{HP(-)H^T + R]^{-1}[I + HP(-)H^TR^{-1}] - R^{-1}\}H$$

$$= I - P(-)H^T\{[HP(-)H^T + R]^{-1}[R + HP(-)H^T]R^{-1} - R^{-1}\}H$$

$$= I - P(-)H^T\{IR^{-1} - R^{-1}\}H$$

$$= I$$

and we have verified equation 9)

Now we derive the alternative form for K_k as given in equation 7). Starting with equation 5) for K_k and once again,

⑥

dropping subscripts, we have

$$K = P(-)H^T[HP(-)H^T+R]^{-1}$$

$$= \{I\}P(-)H^T[HP(-)H^T+R]^{-1}$$

$$= \{P(+)[P^{-1}(+)]\}P(-)H^T[HP(-)H^T+R]^{-1}$$

$$= \{P(+)[P^{-1}(-)+H^TR^{-1}H]\}P(-)H^T[HP(-)H^T+R]^{-1}$$

where we have substituted equation 9) for $P^{-1}(+)$ and the chain continues as

$$K = P(+)\{IH^T+H^TR^{-1}HP(-)H^T\}[HP(-)H^T+R]^{-1}$$

$$= P(+)H^T\{I+R^{-1}HP(-)H^T\}[HP(-)H^T+R]^{-1}$$

$$= P(+)H^TR^{-1}\{R+HP(-)H^T\}[HP(-)H^T+R]^{-1}$$

$$= P(+)H^TR^{-1}$$

This then completes the work necessary to write 8) as follows

(1)

$$\hat{X}_k(+) = \hat{X}_k(-) + P_k(+) H_k^T R_k^{-1} [z_k - H_k \hat{x}_k(-)]$$

Since all subscripts are k we drop them and write the above as

11) $$\hat{X}(+) = \hat{X}(-) + P(+) H^T R^{-1} [z - H \hat{x}(-)]$$

We shall now develop an exactly similar equation in the least-squares formulation where we iterate in the sense that we start with an original measurement set and then an additional measurement set becomes available.

(8)

Let us assume that the set of l measurements, z, can be expressed as a linear combination of the n elements of a constant vector x plus a random, additive measurement error, v. That is the measurement process is modeled as

12) $$z = Hx + v$$

where z is an $l \times 1$ vector, x is an $n \times 1$ vector, H is an $l \times n$ matrix and v is an $l \times 1$ vector. For $l > n$ the measurement set contains redundant information. In least-squares estimation, one chooses as $\hat{x}$ the value which minimizes the sum of the squares of the deviations (4

$z_i - \hat{z}_i$, i.e., in the case where $l = 3$ and $n = 2$, where

$$z_1 = h_{11} x_1 + h_{12} x_2 + v_1$$

13) $$z_2 = h_{21} x_1 + h_{22} x_2 + v_2$$

$$z_3 = h_{31} x_1 + h_{32} x_2 + v_3$$

we are interested in find $\hat{x} = (\hat{x}_1, \hat{x}_2)$ which minimizes $\sum_{i=1}^{3} v_i^2$, i.e., which minimizes

14) $$v^T v = \sum_{i=1}^{3} \left(z_i - \sum_{j=1}^{2} h_{ij} \hat{x}_j \right)^2$$

or in another form, we are find $\hat{x}$ which minimizes

15) $$J = (z - H\hat{x})^T (z - H\hat{x})$$

(10)

The above assumes equal weights for each of the terms

16) $$V_i^2 = \left(z_i - \sum_{j=1}^{2} h_{ij} \hat{x}_j\right)^2$$

We generalize this by using a weighting matrix R^{-1} and minimize

17) $$J = V^T R^{-1} V$$

For simplicity let us assume that R^{-1} is a diagonal matrix $\{R_{ii}\}$ and hence we want to find $\hat{x}_1$ and $\hat{x}_2$ which minimize

18) $$J = \sum_{i=1}^{3} R_{ii} \left(z_i - \sum_{j=1}^{2} h_{ij} x_j\right)^2$$

Differentiating with respect to x_1 and x_2 we get

$$19)\quad \begin{cases} \dfrac{\partial J}{\partial x_1} = 2\displaystyle\sum_{i=1}^{3} r_{ii}\Big(z_i - \sum_{j=1}^{2} h_{ij}\, x_j\Big) h_{i1} \\[2ex] \dfrac{\partial J}{\partial x_2} = 2\displaystyle\sum_{i=1}^{3} r_{ii}\Big(z_i - \sum_{j=1}^{2} h_{ij}\, x_j\Big) h_{i2} \end{cases}$$

Setting the above derivatives equal to zero to find the values of x_1 and x_2, namely $\hat{x}_1$ and $\hat{x}_2$ that minimize J we get

$$20)\quad \begin{cases} \Big(\displaystyle\sum_{i=1}^{3} h_{i1}\, r_{ii}\, h_{i1}\Big)\hat{x}_1 + \Big(\displaystyle\sum_{i=1}^{3} h_{i1}\, r_{ii}\, h_{i2}\Big)\hat{x}_2 = \displaystyle\sum_{i=1}^{3} h_{i1}\, r_{ii}\, z_i \\[2ex] \Big(\displaystyle\sum_{i=1}^{3} h_{i2}\, r_{ii}\, h_{i1}\Big)\hat{x}_1 + \Big(\displaystyle\sum_{i=1}^{3} h_{i2}\, r_{ii}\, h_{i2}\Big)\hat{x}_2 = \displaystyle\sum_{i=1}^{3} h_{i2}\, r_{ii}\, z_i \end{cases}$$

In vector-matrix notation we can write 20) as

$$21)\quad H^T R^{-1} H \hat{x} = H R^{-1} z$$

and solving for $\hat{x}$ we get

$$\hat{x} = (H^T R^{-1} H)^{-1} H R^{-1} z .$$

Now for iterated least-squares. (12)

Let $\hat{x}(-)$ be the weighted least-square estimate corresponding to an original linear measurement set z_0 with measurement matrix H_0 ($z_0 = H_0 x + v$) and weighting matrix R_0^{-1}. By analogy with the above equation 22) we have

23) $$\hat{x}(-) = (H_0^T R_0^{-1} H_0)^{-1} H_0^T R_0^{-1} z_0$$

Now an additional measurement set z becomes available. Defining the following for the complete measurement set

(13)

24) $H_1 = \begin{bmatrix} H_0 \\ H \end{bmatrix}$, $z_1 = \begin{bmatrix} z_0 \\ z \end{bmatrix}$, $R_1 = \left[\begin{array}{c|c} R_0 & 0 \\ 0 & R \end{array}\right]$

the new estimate, $\hat{x}(+)$, is

25) $$\hat{x}(+) = (H_1^T R_1^{-1} H_1)^{-1} H_1^T R_1^{-1} z$$

Now

26) $$H_1^T R_1^{-1} H_1 = [H_0^T, H^T] \left[\begin{array}{c|c} R_0^{-1} & 0 \\ 0 & R^{-1} \end{array}\right] \begin{bmatrix} H_0 \\ H \end{bmatrix}$$

$$= [H_0^T R_0^{-1}, H^T R^{-1}] \begin{bmatrix} H_0 \\ H \end{bmatrix}$$

$$= H_0^T R_0^{-1} H_0 + H^T R^{-1} H$$

and

27) $$H_1^T R_1^{-1} z_1 = [H_0^T, H^T] \left[\begin{array}{c|c} R_0^{-1} & 0 \\ 0 & R^{-1} \end{array}\right] \begin{bmatrix} z_0 \\ z \end{bmatrix}$$

$$= [H_0^T R_0^{-1}, H^T R^{-1}] \begin{bmatrix} z_0 \\ z \end{bmatrix}$$

$$= H_0^T R_0^{-1} z_0 + H^T R^{-1} z$$

Hence we have for $\hat{x}(+)$

28) $$\hat{x}(+) = (H_0^T R_0^{-1} H_0 + H^T R^{-1} H)^{-1} (H_0^T R_0^{-1} z_0 + H^T R^{-1} z)$$

Defining

29) $$P^{-1}(-) = H_0^T R_0^{-1} H_0$$

we thus have

30) $$P^{-1}(+) = H_1^T R_1^{-1} H_1$$

$$= [H_0^T, H^T] \left[\begin{array}{c|c} R_0^{-1} & 0 \\ \hline 0 & R^{-1} \end{array}\right] \begin{bmatrix} H_0 \\ H \end{bmatrix}$$

$$= [H_0^T R_0^{-1}, H^T R^{-1}] \begin{bmatrix} H_0 \\ H \end{bmatrix}$$

$$= H_0^T R_0^{-1} H_0 + H^T R^{-1} H$$

$$= P^{-1}(-) + H^T R^{-1} H$$

where we notice the similarity with 9)

Hence from 28) we have

31) $$\hat{x}(+) = [P^{-1}(-) + H^T R H]^{-1} (H_0^T R_0^{-1} z_0 + H^T R^{-1} z)$$

$$= P(+) (H_0^T R_0^{-1} z_0 + H^T R^{-1} z)$$

using 29) and 30)

Using equation 23) for $\hat{x}(-)$ we have (15)

$$H_0 \hat{x}(-) = H_0 (H_0^T R_0^{-1} H_0)^{-1} H_0^T R_0^{-1} z_0$$

$$= H_0 H_0^{-1} R_0 (H_0^T)^{-1} H_0^T R_0^{-1} z_0$$

$$= I R_0 I R_0^{-1} z_0$$

$$= z_0$$

we can write 31) as

32) $$\hat{x}(+) = P(+)(H_0^T R_0^{-1} H_0 \hat{x}(-) + H^T R^{-1} z)$$

$$= P(+)[H_0^T R_0^{-1} H_0 + H^T R^{-1} H] \hat{x}(-)$$

$$+ P(+) H^T R^{-1} [z - H \hat{x}(-)]$$

where we have added and subtracted

$$P(+) H^T R^{-1} H \hat{x}(-)$$

Using 29) and 30) we can write

32) as

$$\hat{x}(+) = P(+)P^{-1}(+)\hat{x}(-) + P(+)H^T R^{-1}[z - H\hat{x}(-)]$$

$$= \hat{x}(-) + P(+)H^T R^{-1}[z - H\hat{x}(-)]$$

which is exactly similar to equation 11) and explains why the Kalman Filter is sometimes referred to as iterated least-squares.

The above material is taken from the book "Applied Optimal Estimation" edited by Arthur Gelb with an assist from Brogans book "Modern Control Theory".

Appendix

(Peter Maybeck, Stochastic Models, Estimation and Control, Vol 1)

Consider a one-dimensional variable x where we have two measurements of x, z_1 and z_2 such that

$$z_1 = x + v_1 \quad \text{and} \quad z_2 = x + v_2$$

Assume that we have no a-priori information about x and that v_1 and v_2 can be modeled as zero-mean Gaussian random variables with variances $\sigma_{z_1}^2$ and $\sigma_{z_2}^2$, respectively, and that x, v_1, and v_2 are independent random variables.

One way of solving for the best estimate of position would be to consider z_1 and the variance $\sigma_{z_1}^2$ as the "a priori" information about x before the second measurement is taken. This is a sequential (or recursive) estimation procedure. Thus, we use this a priori knowledge to describe the random variable x as a Gaussian random variable with mean z_1 and variance $\sigma_{z_1}^2$.

$(\hat{x}^- = z_1, \ P^- = \sigma_{z_1}^2)$

(18)

Consequently, we consider z_2 as the available measurement to be incorporated into the estimate of position. Since we model z_2 as $x + v$ with v_2 a zero-mean Gaussian variable with variance $\sigma_{z_2}^2$, we have $R = \sigma_{z_2}^2$. We use the formulas

$$P^+ = \left[(P^-)^{-1} + H^T R^{-1} H\right]^{-1}$$

$$K = P^+ H^T R^{-1}$$

$$\hat{x}^+ = \hat{x}^- + K\left[z - H\hat{x}^-\right]$$

where

$$P = \sigma_{z_1}^2$$

$H = 1$ and thus $H^T = 1$

$R = \sigma_{z_2}^2$ and thus $R^{-1} = \dfrac{1}{\sigma_{z_2}^2}$

$$\hat{x} = z_1$$

Hence

$$P^+ = \left[\frac{1}{\sigma_{z_1}^2} + (1)\left(\frac{1}{\sigma_{z_2}^2}\right)(1)\right]^{-1} = \frac{\sigma_{z_1}^2 \sigma_{z_2}^2}{\sigma_{z_1}^2 + \sigma_{z_2}^2}$$

$$K = \frac{\sigma_{z_1}^2 \sigma_{z_2}^2}{\sigma_{z_1}^2 + \sigma_{z_2}^2} (1) \frac{1}{\sigma_{z_2}^2} = \frac{\sigma_{z_1}^2}{\sigma_{z_1}^2 + \sigma_{z_2}^2}$$

$$\hat{x} = z_1 + \frac{\sigma_{z_1}^2}{\sigma_{z_1}^2 + \sigma_{z_2}^2}\left(z_2 - (1)z_1\right) = \frac{\sigma_{z_2}^2}{\sigma_{z_1}^2 + \sigma_{z_2}^2} z_1 + \frac{\sigma_{z_1}^2}{\sigma_{z_1}^2 + \sigma_{z_2}^2} z$$

(19)

Another way of solving for the best estimate of position is to assume no a priori information about x, and to incorporate the two measurements simultaneously, i.e. in a batch mode. If there is no a priori information, we could model this through a Gaussian random variable with infinite variance, $P^{-1} = \infty$, or equivalently, $(P^-)^{-1} = 0$. The measurement is

$$z = \begin{pmatrix} z_1 \\ z_2 \end{pmatrix} = \begin{pmatrix} 1 \\ 1 \end{pmatrix} x + \begin{pmatrix} v_1 \\ v_2 \end{pmatrix} = H x + v$$

where v is modeled as a zero-mean Gaussian noise of covariance R:

$$R = E\{v v^T\} = \begin{bmatrix} E\{v_1^2\} & E\{v_1 v_2\} \\ E\{v_1 v_2\} & E\{v_2^2\} \end{bmatrix} = \begin{bmatrix} \sigma_{z_1}^2 & 0 \\ 0 & \sigma_{z_2}^2 \end{bmatrix}$$

where the off diagonal zeros are due to v_1 and v_2 being independent and thus uncorrelated.

Again using the formulas

$$P^+ = [(P^-)^{-1} + H^T R^{-1} H]^{-1}$$

$$K = P^+ H^T R^{-1}$$

$$\hat{x}^+ = \hat{x}^- + K[z - H \hat{x}^-]$$
$$= [I - KH]\hat{x} + Kz$$

where now

$$P^- = \infty, \quad (P^-)^{-1} = 0$$

$$H = \begin{pmatrix} 1 \\ 1 \end{pmatrix} \text{ and thus } H^T = (1, 1)$$

$$R = \begin{pmatrix} \sigma_{z_1}^2 & 0 \\ 0 & \sigma_{z_1}^2 \end{pmatrix} \text{ and thus } R^{-1} = \begin{pmatrix} \frac{1}{\sigma_{z_1}^2} & 0 \\ 0 & \frac{1}{\sigma_{z_2}^2} \end{pmatrix}$$

and we ignore $\hat{x}^-$ temporarily

Hence $$P^+ = \left[0 + (1,1) \begin{pmatrix} \frac{1}{\sigma_{z_1}^2} & 0 \\ 0 & \frac{1}{\sigma_{z_2}^2} \end{pmatrix} \begin{pmatrix} 1 \\ 1 \end{pmatrix} \right]^{-1} = \frac{\sigma_{z_1}^2 \sigma_{z_2}^2}{\sigma_{z_1}^2 + \sigma_{z_2}^2}$$

$$K = \frac{\sigma_{z_1}^2 \sigma_{z_2}^2}{\sigma_{z_1}^2 + \sigma_{z_2}^2} (1,1) \begin{pmatrix} \frac{1}{\sigma_{z_1}^2} & 0 \\ 0 & \frac{1}{\sigma_{z_2}^2} \end{pmatrix} = \left(\frac{\sigma_{z_2}^2}{\sigma_{z_1}^2 + \sigma_{z_2}^2}, \frac{\sigma_{z_1}^2}{\sigma_{z_1}^2 + \sigma_{z_2}^2} \right)$$

Now $$KH = \left(\frac{\sigma_{z_2}^2}{\sigma_{z_1}^2 + \sigma_{z_2}^2}, \frac{\sigma_{z_1}^2}{\sigma_{z_1}^2 + \sigma_{z_2}^2} \right) \begin{pmatrix} 1 \\ 1 \end{pmatrix} = \frac{\sigma_{z_1}^2 + \sigma_{z_2}^2}{\sigma_{z_1}^2 + \sigma_{z_2}^2} = 1$$

Hence $I - KH = 1 - 1 = 0$

and we can ignore $\hat{x}^-$ permanently since it is multiplied by $I - KH$ and the formula for $\hat{x}^+$ reduces to

$$\hat{x}^+ = Kz = \left(\frac{\sigma_{z_2}^2}{\sigma_{z_1}^2 + \sigma_{z_2}^2}, \frac{\sigma_{z_1}^2}{\sigma_{z_1}^2 + \sigma_{z_2}^2} \right) \begin{pmatrix} z_1 \\ z_2 \end{pmatrix}$$

$$= \frac{\sigma_{z_2}^2}{\sigma_{z_1}^2 + \sigma_{z_2}^2} z_1 + \frac{\sigma_{z_1}^2}{\sigma_{z_1}^2 + \sigma_{z_2}^2} z_2 \text{ as before.}$$